Lecture Notes in Artificial Intelligence   6202

Edited by R. Goebel, J. Siekmann, and W. Wahlster

Subseries of Lecture Notes in Computer Science

Michael May   Lorenza Saitta (Eds.)

Ubiquitous Knowledge Discovery

Challenges, Techniques, Applications

 Springer

Series Editors

Randy Goebel, University of Alberta, Edmonton, Canada
Jörg Siekmann, University of Saarland, Saarbrücken, Germany
Wolfgang Wahlster, DFKI and University of Saarland, Saarbrücken, Germany

Volume Editors

Michael May
Fraunhofer IAIS
Schloss Birlinghoven, 53754 Sankt Augustin, Germany
E-mail: michael.may@iais.fraunhofer.de

Lorenza Saitta
Università del Piemonte Orientale Amedeo Avogadro
Dipartimento di Informatica
Viale Teresa Michel 11, 13100, Alessandria, Italy
E-mail: saitta@di.unipmn.it

Library of Congress Control Number: 2010936640

CR Subject Classification (1998): I.2.4, I.2, H.3-4, C.2, F.1, H.2.8

LNCS Sublibrary: SL 7 – Artificial Intelligence

ISSN   0302-9743
ISBN-10   3-642-16391-2 Springer Berlin Heidelberg New York
ISBN-13   978-3-642-16391-3 Springer Berlin Heidelberg New York

springer.com

© Springer-Verlag Berlin Heidelberg 2010
Printed in Germany

Typesetting: Camera-ready by author, data conversion by Scientific Publishing Services, Chennai, India
Printed on acid-free paper   06/3180

Preface

Over the last years, ubiquitous computing has started to create a new world of small, heterogeneous, and distributed devices that have the ability to sense, to communicate and interact in ad hoc or sensor networks and peer-to-peer systems. These large-scale distributed systems, in many cases, have to interact in real-time with their users. Knowledge discovery in ubiquitous environments (KDubiq) is an emerging area of research at the intersection of the two major challenges of highly distributed and mobile systems and advanced knowledge discovery systems. It aims to provide a unifying framework for systematically investigating the mutual dependencies of otherwise quite unrelated technologies employed in building next-generation intelligent systems: machine learning, data mining, sensor networks, grids, peer-to-peer networks, data stream mining, activity recognition, Web 2.0, privacy, user modeling and others.

In a fully ubiquitous setting, the learning typically takes place *in situ*, inside the small devices. Its characteristics are quite different from currently mainstream data mining and machine learning. Instead of offline-learning in a batch setting, sequential learning, anytime learning, real-time learning, online learning, etc.—under real-time constraints from ubiquitous and distributed data— is needed. Instead of learning from stationary distributions, concept drift (the change of a distribution over time) is the rule rather than the exception. Instead of large stand-alone workstations, learning takes place in unreliable, highly resource constrained environments in terms of battery power and bandwidth.

To explore this emerging field of research, a networking project has been funded since 2006 by the European Commission under grant IST-FP6-021321[1]: KDubiq (knowledge discovery in ubiquitous environments) is a coordination action at the intersection of the two major challenges of highly distributed and mobile systems and advanced knowledge discovery systems. A basic assumption of the project is that what seems to be a bewildering array of different methodologies and approaches for building "smart," "adaptive," "intelligent" ubiquitous knowledge discovery systems can be cast into a coherent, integrated set of key areas centered on the notion of learning from experience. The objective of KDubiq is to provide this common perspective, and to shape a new area of research. For doing so, the KDubiq coordination action has coordinated relevant research done on learning in many subfields, including:

- machine learning and statistics
- knowledge discovery in databases or data mining
- distributed and embedded computing
- mobile computing

---

[1] See the website www.kdubiq.org for details about the project.

- human computer interaction (HCI)
- cognitive science

A major goal was to create for the first time a forum to bring these individual research lines together, to consolidate the results that have already been achieved, and to pave the way for future research and innovative applications. For doing so, KDubiq has organized a large number of workshops, summer schools, tutorials and dissemination events to bring together this new community.[2] One important means to focus the activities and discussions was a collaborative effort to provide *a blueprint for the design of ubiquitous knowledge discovery systems.* A number of working groups on relevant topics have been established. Their goal was to create a conceptual framework for this new line of research, to survey the state of the art, and to identify future challenges, both on the theoretical and the applications side.

The result of this collaborative effort is Part I of this book. This blueprint manifests the vision and serves as a practical guide for further, integrated advances in this field, towards, in the long-term, building truly autonomous intelligent systems.

Overview of the Book. Part I of the book aims to provide a conceptual foundation for the new field of ubiquitous knowledge discovery, discussing the state of the art, highlighting challenges and problems, and proposing future directions. Although at some points technical examples are given for illustration, the aim of this chapter is rather on the non-technical, conceptual side.

While Part I is divided into individually authored chapters, it should be seen as a collaborative effort by the working groups of the KDubiq coordination action. Each chapter was read and commented by the other working group members, and influenced by the discussions and findings of the individual working groups.[3] Hence, the chapters should be seen as an integrated whole.

Part I of the book is structured as follows. Chapter 1 gives an introduction to the topic and the fundamental issues. Chapter 2 provides an overview on three distributed infrastructures for ubiquitous computing. Chapter 3 discusses how the learning setting itself changes in a ubiquitous environment, when compared to a traditional learning set-up. Chapter 4 defines general characteristics of data in ubiquitous environments. Chapter 5 takes up the issues of privacy and security, arguing that they are critical for the deployment and user acceptance of KDubiq systems. Chapter 6 is devoted to the human-centric view of ubiquitous knowledge discovery systems. Finally, Chapter 7 contains a collection of potential application areas for KDubiq, providing pointers to the state of the art, to existing applications (if available) and to challenges for future research.

---

[2] See http://www.kdubiq.org/kdubiq/images/KDubiq.newsletter.pdf for details.

[3] In some cases, project partners provided input for some sections, but not for the whole chapter. Where this is the case, it is stated in the footnotes of individual sections.

Part II contains selected approaches to ubiquitous knowledge discovery and treats specific aspects in detail. The contributions have been carefully selected to provide illustrations and in-depth discussions for some of the major findings of Part I.

The contribution by Antoine Cornéujols takes up in greater detail two fundamental challenges for learning in ubiquitous environments: incrementality and non-stationarity. The chapter by Severo and Gama investigates change detection in temporal data, when the assumption of stationarity is not met. Sharfman, Schuster, and Keren have the topic of monitoring changes in highly distributed data stream systems using a geometric approach. The contribution by Inan and Saygin addresses another highly important dimension of ubiquitous knowledge discovery: privacy. It discusses the problem of clustering horizontally partitioned spatio-temporal data (the data sets at the nodes have the same attributes, but different instances) in a privacy-preserving manner. The chapter by Katharina Morik describes a peer-to-peer Web 2.0 application for collaborative structuring of multimedia collections. The chapter by Rasmus Pedersen broadens the discussion and provides an overview on the topic of learning in micro-information systems. The final chapter by Hillol Kargupta and co-workers describes the MineFleet system, one of the few commercially available ubiquitous knowledge discovery systems. Coordinating a network with more than 50 partner institutions and several hundred individual members is a complex, and sometimes daunting, task. We thank the many researchers and practitioners that contributed to the discussions that led to this book in various ways[4]; the invited speakers that helped us to sharpen our understanding of the research issues involved; the numerous workshop and summer school attendees; the project reviewers for constructive criticism; and the EC project officers Fabrizio Sestini and Paul Hearn for their support. We also thank Tino Sanchez for maintaining the project website. By far the greatest thanks, however, go to Ina Lauth. She has coordinated the network activities for three years and did a superb job in making it both a vibrant and pleasant experience for everyone, thereby having a great share in the successful outcome of the project.

The preparation of this book has been supported by the European Commission under grant KDubiq, IST-FP6-021321, which we gratefully acknowledge. We hope that the reader will agree with us that ubiquitous knowledge discovery in many ways holds potential for radically changing the way machine learning and data mining is done today, and share our excitement about this new field of research.

June 2010   Michael May
                                                                  Lorenza Saitta

---

[4] They are too numerous to list individually here, but see the KDubiq newsletter, the members page on KDubiq.org, and the acknowledgments for individual chapters of Part I.

Table of Contents

Part I

A Blueprint for Ubiquitous Knowledge Discovery

Introduction: The Challenge of Ubiquitous Knowledge Discovery

Michael May and Lorenza Saitta

[1] Fraunhofer Institut Intelligente Analyse- und Informationssysteme,
Department Knowledge Discovery,
Sankt Augustin, Germany
[2] Universitá Degli Studi Del Piemonte Orientale Amedeo Avogadro,
Alessandria, Italy
michael.may@iais.fraunhofer.de, saitta@di.unipmn.it

1 The Object of Investigation: Ubiquitous Knowledge Discovery

In the past, the development of machine learning approaches was to some extent motivated by the availability of data and increased computational power. Ubiquitous computing bears the promise of stimulating a similar leap forward. Small devices can now be installed in many places, mobile and wearable devices enable registration of large amounts of information, thus generating a wide range of new types of data for which new learning and discovery methods are needed, far beyond existing ones.

Hence, Ubiquitous Knowledge Discovery (KDubiq)[3] focuses on the extension of data mining to *modern computing*. Notable breakthroughs have revolutionized computer science in the last decade, along several perspectives: hardware, communication/networks, and usage. Gradually, the software world becomes more aware that new algorithmic paradigms are required to make the most of new architectures, to handle new demands, and to sail towards New Intelligence Frontiers. KDubiq is meant to favour the emergence of such new algorithmic paradigms in the domain of data mining and machine learning.

Resting on the above grounds, ubiquitous knowledge discovery is then a *new discipline* that has its roots in the parent fields of data mining, on the one hand, and ubiquitous computing, on the other. The novelty of the discipline stems from the timely co-occurrence and synergetic integration among the new needs from a multitude of new types of users, the capillary distribution of data and information, and the new resources provided by today's computational environments.

Relevant research is done on learning in many subfields, including:

- machine learning and statistics
- knowledge discovery in databases or data mining
- distributed and embedded computing
- mobile computing
- human computer interaction (HCI)
- cognitive science.

M. May and L. Saitta (Eds.): Ubiquitous Knowledge Discovery, LNAI 6202, pp. 3–18, 2010.
© Springer-Verlag Berlin Heidelberg 2010

The *object of investigation* of KDubiq encompasses the whole process of knowledge discovery in mobile, finely distributed, interacting, dynamic environments, in presence of massive amounts of heterogeneous, spatially and temporally distributed sources of data. More precisely, KDubiq aims at developing algorithms and systems for:

- *Locating data sources* - The current availability of world-wide distributed databases, Web repositories, and sensor networks adds a novel problem to todays knowledge discovery, i.e., the need to locate, in a transparent way, possible sources of data required to satisfy a user's demand.
- *Rating data sources* - As multiple sources may usually be identified as candidate data suppliers, effective evaluation methods must be designed to rate the sources w.r.t. the search target.
- *Retrieving information* - The nature and type of data available today are different from the past, as much emphasis is set on semantically rich, multimodal, interacting pieces of information, which need clever and fast data fusion processes.
- *Elaborating information* - The development of new approaches and algorithms must match the changes in the data distribution and nature. Algorithms are not only faced to the scalability problem (as in the past), but also to resource constraints, communication needs, and, often, real-time processing. Moving the data mining software from a central database to the ubiquitous computing devices demands a design of systems and algorithms which differs considerably from that of classical data mining. This is the price to pay for introducing intelligence in small devices.
- *Personalizing the discovery process* - Due to the large number and variety of potential users of KDubiq, adaptation to individual users is a must. People, in fact, more than ever play a pivotal role in the knowledge handling process: people create data, data and knowledge are about people, and people are often the ultimate beneficiaries of the discovered knowledge. This requires that the benefit-harm tradeoff for different stakeholder groups is understood and used for system design.
- *Presenting results* - As most often people are the intended users of the discovered knowledge, it is very important that adequate visualization techniques be designed to easy their task of interpreting the (intermediate and final) results of the process. User models may help designing more effective human/machine interactions, and guiding the development of new services.

Even though not all KDubiq problems are concerned with all of the above aspects, each one of them needs substantial improvements over existing techniques, when not a totally new re-thinking, offering thus occasion for advanced research.

2   Characteristic Features of KDubiq Research

A possible direction orienting KDubiq research is to be concerned with the matter of scale. Distribution of both data sources and processing devices may become fine-grained, letting the solution of a problem emerge from the cooperation among many agents, each endowed with only small capabilities, but able to interact and communicate with others. This landscape has multiple effects: first of all, the fine-grained computational distribution implies the use of small devices, with possibly limited power. The consequence is a resource-aware approach, which can be better understood by analogy with the notion, well known in artificial intelligence, of bounded rationality vs. plain rationality[1], where the cost of the processing becomes an integral part of the process design itself.

A second important aspect is that, in the development of KDubiq systems, there are external parameters that influence the design choices and were not part of the rational under previous data mining paradigms: for instance, the KDubiq process must pay attention to the origin/location of the data, to the willingness of people to share data and to assess data reliability, to the specific nature of the computational means (hardware-aware algorithms), and to the user's expectations (user modeling) and profile (personalized interestingness criteria). In addition, all of this must cope with *privacy* and *security* issues. In fact, the role of humans in the process is increased: not only privacy is more relevant, but also *trust*, especially in handling data coming from social networks, and in peer-to-peer interactions.

A third aspect of KDubiq is *embedded* processing. Instead of collecting (possibly distributed) data, elaborating them in one or few processing sites and routing the results to the users, the elaboration process can be transparently embedded in devices which answer to local needs *in situ* using local information, while communicating with others to face more general requests. Examples can be found in surveillance cameras, wearable health sensors, cell phones, and many others. Embedding is a side-effect of the fine-grained scale of the processing units distribution.

Finally, the types of data to be handled may show extensive variations over *time* and *space*, requiring analysis methods able to cope with possibly stringent *real-time* processing. Regarding variations, there is yet another aspect to consider, namely the possibility of erratic behaviour, due to a large spectrum of possible causes, for instance system crashes. If erratic behaviour may occur, algorithms are strongly influenced, because they must be robust in face of it.

In summary, KDubiq research is characterized by (1) the presence of a population of agents (small devices) provided with computing capabilities, (2) a rich semantics of the data, reflecting the behavior of the agents and their interactions, (3) an extreme distribution of such data, possibly with no chance of centralization, even partial, (4) a continuous flows of incoming data, (5) a complex scenario for the management of security and privacy, and (6) a deep involvement of people as producers of data and consumers of results.

---

[1] The notion of bounded rationality has been introduced by Herbert Simon, to revise the assumption, common in social and economic sciences, that humans could be approximated by rational reasoning entities. Bounded rationality accounts for the fact that perfectly rational decisions are often not feasible in practice due to the shortage of computational resources available.

3   Differences between KDubiq and Distributed Data Mining

Data mining and KDubiq share the same goal: extracting interesting and useful knowledge from various types of data. As an analogy, let us mention the field of programming languages: object-oriented programming is part of the field, but one that has introduced a new and influential programming paradigm indeed. In the same way, KDubiq is part of data mining, especially close to distributed data mining (DDM), which, in some sense, KDubiq brings to a somewhat extreme realization.

At least four fundamental tracts play a role in differentiating KDubiq from DDM:

- *Scale* - In DDM, distribution is usually medium scale: a few data sources (databases, Web archive, sensors signals, ...) are exploited for the discovery process and/or a few computer facilities elaborate the data, leaving to a further step the collection of the results and its presentation to the user. In KDubiq, the number of explored data sources may be of orders of magnitude larger (even in the thousands or more). This may happen, for example, when a large part of the WWW has to be explored, or the input comes from very large sensor networks (electrical power net, grid systems, traffic sensors, weather parameter sensors, ...). This level of distribution requires approaches that are qualitatively, and not only quantitatively, different from existing ones.
- *Communication* - Given the fine-grained distribution of the computing units, it is not thinkable to set up a master computer that collects the individual results and put them together for the user. On the contrary, the final result shall emerge as a whole from the network of communications among the local computing units.
- *Resource-awareness* - The ubiquity of processing is usually coupled, as already mentioned, with a reduction of the computing power of the involved devices. Moreover, the discovery task may be requested to work in real-time. Again, performing a task under these operating conditions requires that novel approaches and methodologies be invented.
- *Integration* - Finally, the whole process of discovery cannot be cleanly split into its component phases (data collection, data cleaning, pre-processing, algorithm application, interpretation, visualization), but all these phases must interact, as they are intertwined into the computing agent network.

4   Why Are Current Data Mining Techniques Insufficient?

Given the description of KDubiq's characteristic features, it appears quite obvious why current DDM methods are insufficient to meet the challenges set by the ubiquity of the discovery task.

Current methods are unable to cope with a finely distributed fleet of small agents and with their interactions. Todays degree of computation distribution does not consider communication among agents, and aggregation of results is always done in an almost centralized manner. In contrast, the extreme distribution of local computation calls for new ways of information exchange and aggregation of partial results: possibly never, during the discovery task, a global view of the whole process is assembled in some place; nevertheless, the user must receive coherent answers within reasonable time delays.

When the input data are generated by a sensor network, the difficulty of coordinating the computation among the agents increases, because the appropriate pieces of data must be routed towards the agent that needs them, without disrupting the semantics of the data and their spatio-temporal patterns. There are currently few algorithms able to do this, especially in real-time, in dynamic environments.

Today, data mining exploits the computational power of machines in an unbounded way, that is, limits are set by the computational time complexity of the algorithms, and not by the exhaustion of the resources. If small devices are used and must respond in an intelligent way, most of the current algorithms are no more applicable, because they will exhaust the computational resources very rapidly. Moreover, the new, resource-aware algorithms that are needed must be able to perform their task with a very limited horizon on the whole problem, relying on little local information but on a lot of communication. As a consequence, current learning and discovery algorithms are too sophisticated for the new generation computing units envisaged by KDubiq.

Even though privacy-aware methods already exist, they are inadequate in a finely distributed environment, where no centralized control for accessing the data exists. Then, more stringent privacy and security constraints arise, in order to regulate access rights of agents, to protect their anonymity and to manage trust and confidence of their interactions.

Finally, the way people interact with KDubiq systems are to be reconsidered. For instance, should cell phones or palm computers become the privileged tools for people to do data mining or enter their own data, then new interfaces are to be designed to cope with the limitation of the physical interaction space.

5   A Cognitive and People-Centric Perspective

From a people-centric perspective, KDubiq means rethinking who is the "user" of a KDubiq system and how all the advances and dimensions of ubiquity affect users of such systems (and other stakeholders). This is not only a matter of interfaces but a perspective that spans the whole design process, from the conception of a system through to its deployment.

KDubiq must work "in real time", "in real environments", and therefore "under conditions of real resource constraints" (such as bounded rationality). Ubiquity enlarges the definition of "users" of a system to a wider range of people encompassing users, stakeholders and communities. Consequently, methods to elicit possibly conflicting interests of the target population are required. Conflicts arise mainly due to data coming from different users or stakeholders, obtained from different data/knowledge sources. As conflicts should not be avoided, methods for dealing with them have also to be developed.

Data collection from ubiquitous users must cope with two major problems: challenges of obtaining data and challenges of their representativeness. Users differ in their ability to provide information as well as in their willingness to share it. Differences in privacy issues and willingness to share information ask for a detailed examination of the extent to which data gathering would constitute an intrusion into the private space. Furthermore, a user's background affects the way opinions are expressed. This needs to be taken into account either through a culturally adapted conception of data gathering tools or through appropriate data processing that considers these differences.

Ubiquity of people leads to heterogeneous data sets due to different contexts. User-centered knowledge discovery hence requires data processing that takes background knowledge about the users and their context into account. Ontologies and 'folksonomies' have been analyzed as a possible solution that needs to be extended to tackle the above mentioned context problem. Given the increasing amount of multilingual data sets, knowledge discovery should also take into consideration research results regarding multilingual information retrieval tools.

6 Example of Typical KDubiq Applications

Even though the spectrum of potential application fields is really very large, we have selected here two of them, that can be used to exemplify "typical" KDubiq applications: one is related to activity discovery, and the second one to providing a web service for music file sharing.

6.1 Activity Recognition – Inferring Transportation Routines from GPS-Data

The widespread use of GPS devices has led to an explosive interest in spatial data. Classical applications are car navigation and location tracking. Intensive activity, notably in the ubiquitous computing community, is underway to explore additional application scenarios, e.g., in assistive technologies or in building models of the mobile behavior of citizens, useful for various areas, including social research, planning purposes, and market research.

We discuss an application from assistive technologies, analyze its strength and shortcomings and identify research challenges from a KDubiq perspective.

The *OpportunityKnocks* prototype [1] consists of a mobile device equipped with GPS and connected to a server. An inference module running on the server is able to learn a person's transportation routines from the GPS data collected. It is able to give advice to persons, e.g., which route to take or where to get off a bus, and it can warn the user in case he commits errors, e.g., takes the wrong bus line. The purpose of the system is to assist cognitively impaired persons in finding their way through city traffic.

This application meets the main criteria for ubiquitous systems: the device is an object moving in space and time, it is operating in an unstable and unknown environment; it has computing power, and has a local view of its environment only; it reacts in real-time and it is equipped with GPS-sensors and exchanges information with other objects (e.g., satellites, the server). Since both the environment and the behavioral patterns are not known in advance, it is impossible to solve this task without the system being able to learn from a user's past behavior. Thus, machine learning algorithms are used to infer likely routes, activities, transportation destinations and deviations from a normal route.

The basic infrastructure of OpportunityKnocks is a client/server setup. A GPS-Device is connected to a mobile phone. The mobile phone can connect to a server via GPRS and transmit the GPS signals. Wireless communication is encrypted. The server analyses the data, utilizing additional information about the street network or bus schedules from the Internet. Using this information the person is located and the system makes inferences about his current behavior and gives suggestions what to do next. This information is sent back to the client and communicated to the user with the help of an audio/visual interface.

Although innovative, the architecture of this prototype will face a number of practical problems:

- When there is no phone signal, communication with the server is impossible, and the person may get lost.
- When there is no reliable GPS signal, e.g., in a train, indoors, in an underground station, guidance is impossible. GPS for pedestrians in an urban environment is unreliable.
- Communicating via a radio network with a server consumes a lot of battery power (the system works only 4 hrs under continuous operation).
- Continuously tracking of a person and centrally collecting the data creates strong privacy threats.

The KDubiq paradigm asks for distributed, intelligent and communicating devices integrating data from various sources. A "KDubiq Upgrade" would result in a much more satisfactory design for the prototype. The upgrade is guided by the following imperatives:

1. Move the machine learning to the mobile device.
2. Add more sensors and learn from them.
3. Let the device learn from other devices.
4. Respect privacy.

With current state-of-the-art mobile phone technology, all technical infrastructures for upgrading are in place. What is missing currently are the proper knowledge discovery tools.

If the major part of the learning is done on the mobile device – especially that part that refers to localization on the street map – there is no need for constant server communication, and assistance becomes more reliable.

Moving the machine learning to the device introduces new tasks and additional criteria in assessing the value of a machine learning algorithm that existing algorithms are not designed to meet. Addressing these criteria results in new algorithms that often would not make sense in a non-KDubiq environment.

To give an example, due to high communication costs, OpportunityKnocks is continuously operating for 4 hrs before the battery is exhausted – not enough for a real-world environment. Since network communication is very expensive and thus a bottleneck (see Chapter 3), *power consumption* appears as an additional criterion for the design of a learning algorithm. In a traditional environment, this criterion is irrelevant.

A solution that calculates *everything* on the mobile device and does not communicate at all will not do. OpportunityKnocks uses Dynamic Bayesian Networks. This is heavy machinery, and doing all computations on the device would be slow; moreover, the computation would require a lot of resources, so that, again, the battery would quickly run low. Instead, a solution is needed that locally computes and pre-aggregates results and communicates only few data via the radio network. Splitting the computation into an energy and computationally efficient online part yielding highly compressed models, transmitting only this compressed information and performing computationally intensive parts on the server is a better solution. Pioneering work in this direction – in the context of vehicle monitoring – has been done by Kargupta [2]. Kargupta shows that for the online part new algorithms are necessary that trade accuracy against efficiency. The specific trade off is dictated by the application context, and the choice made in the vehicle monitoring application would be hard to motivate in an offline-context (or even for the current application).

Once the main part of learning is done on the device, opportunities occur for giving the system access to more sensory and background information, including information generated by other similar devices.

If the system is equipped with a gyroscope and/or accelerometer and has a local map, short-term navigation is possible in the absence of a GPS signal. Using multi-sensor input for online-analysis is a common approach for car navigation already, combining the signals e.g. using Kalman filtering. For mainstream machine learning and data mining, it challenges some basic assumptions, since data arrives not as a batch but in a streaming setting (cf. Chapter 4 for details).

If the device is additionally able to communicate with other devices – e.g. using WiFi or Bluetooth – one use of WiFi could be to complement GPS for indoor localization. Even more interesting would be communication with a collection of *other similar systems*, e.g. to infer the transportation mode by taking into account the local knowledge of these other systems.

The distributed, collaborative nature of KDubiq is thus a source for new algorithms. (cf. Section 6.2 in this chapter). In contrast to almost all approaches to distributed mining, the algorithms can exploit other devices' knowledge without aiming for a *global model*. They yield new and different algorithms that would simply make no sense in a non distributed environment – and therefore did not exist before.

The original scenario creates severe privacy threats. Once the mining goes to the machine, the privacy sensitive data can remain with the person, and the privacy risks of central monitoring are removed.

Additional risks occur however due to the fact that devices communicate with other devices. Thus privacy constraints are a further third source for new algorithms. The task is to distribute the computing in such a way that no computing device not under the control of the user has access to information that infringes privacy. Design options for privacy preserving data mining are discussed in Chapter 6.

To sum up, existing algorithms cannot solve the problem because they have not been designed for this task, and do not take requirements into account that emerge in an ubiquitous context.

A further kind of challenges derives from the ubiquitous computing roots of KDubiq.

It is about careful selection, adaptation and integration of partial solutions, both algorithmic and technical, resulting in a delicate mixture of hardware, software and algorithm design issues. Some of the associated challenges may be best considered as system integration or *engineering challenges.* The challenge is that the sum is qualitatively more than its parts – and exponentially more complex to implement. Often underrated by machine learning purists, for the new field of KDubiq these challenges are at the core. Thus, the interdisciplinary nature of KDubiq asks for a broader view what constitutes a research challenge and to include activities as they are commonly carried out in other research fields such as ubiquitous computing or robotics.[2]

Overall, the challenges for the current application consist in building learning algorithms for distributed, multi-device, multi-sensor environments. While partial suggestions exist on how to implement privacy-preserving, distributed, collaborative algorithms, respectively, there is hardly any existing work that properly addresses all the dimensions at the same time in an integrated manner. *Yet as long as one of these dimensions is left unaddressed, the ubiquitous knowledge discovery prototype will not be fully operational in a real-world environment.* We need both new algorithms – including analysis and proof about their complexity and accuracy – and an engineering approach for integrating the various partial solutions – algorithms, software and hardware – in a working prototype. Building such a prototype would be a challenging topic for a research project.

6.2   Ubiquitous Intelligent Media Organization

With the advent of Web 2.0, collaborative structuring of large collections of multi-media data based on meta-data and media features has become a significant task. While most applications are based on a central server, peer-to-peer solutions start to appear.

---

[2] To prepare the community for this shift in the skill set of a machine learner, KDubiq has organized two summer schools (www.kdubiq.org).

As an example we discuss Nemoz (NEtworked Media Organizer) [4][10][16], which is a Web 2.0-inspired collaborative platform for playing music, browsing, searching and sharing music collections. It works in a loosely-coupled distributed scenario, using P2P technology. Nemoz combines Web 2.0-style tagging, with automatic audio classification using machine learning techniques.

Nemoz is motivated by the observation that a globally correct classification for audio files does not exist, since each user has its own way of structuring the files, reflecting his own preferences and needs. Still, a user can exploit labels provided by other peers as features for his own classification: the fact that Mary, who structures here collection along mood, classifies a song as 'melancholic' might indicate to Bob, who classifies along genre, that it is not a techno song. To support this, Nemoz nodes are able to exchange information about their individual classifications. These added labels are used in a predictive machine learning task. Thus the application is characterized by evolving collections of large amounts of data, scattered across different computing devices that maintain a local view of a collection, exchanging information with other nodes.

Main features of the application are:

1. The application differs from the preceding one in that the (geo)spatio-temporal position of the computing devices does not play an important role; the devices and the media file collections they contain are stored somewhere on some node in the network.
   Yet it is a defining characteristic of the application that two collections C_i and C_j are stored at different places, and it is important whether or not two collections are connected via a neighborhood graph.
2. The computing devices might be connected to a network only temporarily. The collections are evolving dynamically; items are added and deleted, and also classifications can change.
3. The computing nodes have sufficient local computing power to carry out complex tasks, and as long as mobile devices are not included, resource constraints are low.
4. It is a crucial aspect of this application that the nodes maintain a local view, incorporating information from other nodes.
5. The device does do not take autonomous action or actively monitors the state of the collection until some event occurs. Still response-time is an important issue because the application is designed for interactivity.
6. Finally, the distributed nature of the problem is a defining characteristic of the application.

We are not aware of other solutions that are able to automatically learn from other user's classifications while maintaining a local or subjective point of view. This application is a representative of a innovative subclass of applications in a Web 2.0 environment. Whereas most Web 2.0 tagging applications use a central server where all media data and tags are consolidated, the current application is fully distributed.

In many distributed data mining applications, originally centralized data are distributed for improving the efficiency of the analysis.

The current application is different because, firstly, the data are inherently distributed, and secondly, there is no intention to come up with a global model. Communication across peers is used for improving a local classification by feature harvesting. Thus Nemoz introduces a new class of learning problems: the collaborative representation problem and localized alternative cluster ensembles for collaborative structuring (LACE).

From the perspective of KDubiq this is important, since in a non-distributed environment, these new learning scenarios would be very hard to motivate. We see in this case study that ubiquitous knowledge discovery prompts the invention of new learning scenarios – not only problems are solved that could not be solved with existing methods; rather, *new classes of learning problems are invented*.

It is also interesting to note that the extension of the activity recognition scenario could raise a need for similar mechanisms as utilized in Nemoz (see section 6.1). This potential transfer of learning scenarios from two seemingly very unrelated areas – mobile assistive technology and music mining – is made possible by analyzing the applications in a common framework.

Several future extensions of Nemoz seem possible:

- Apart from the need for fast response-times because of interactivity requirements, resource constraints for the mining algorithms are not discussed – probably because experiments are done on notebooks or workstations.
  But once devices such as mobile phones are included (and they are mentioned as potential devices), these considerations will become important. The classification algorithms used are quite efficient (e.g., decision trees) and could be run without difficulties on a mobile device.
- User can mark their tag structures as private or public. Private folders are not visible to other users while browsing. Private folders are, however, used for data mining. It seems by sending a request, an attacker would be able to infer which music files are probably stored on another computer. This reveals information about an user. It would be interesting to find out whether the sharing can be done in a privacy-preserving manner.
- In extension of characteristic (5) above, one can easily imagine e.g. a publish-subscribe scenario where a node broadcasts a message to interested parties if a new item appears in a certain group of his collection. In this additional scenario, temporal, and spatial aspects are very important, and also acting in real-time, e.g. if it is important to be the first to classify a hitherto unknown song.

Thus we can see can see that a further development of Nemoz, addressing more fully the different dimensions of KDubiq (small devices, privacy, a real-time scenario) leads to interesting new research questions.

7 Technical Challenges to Be Overcome by Future Research

Challenges to be overcome by future research involve all components of the data mining process: data, algorithms, and interfaces. They can be roughly summarized by saying that large problems must be solved with many small computational elements.

First of all, mobility must be tamed; as agents move, operate and interact with each other according to some predefined purposes, their movements are to be traced, and the recorded traces are augmented with richer background information: occurrence of events, such as load and delivery of goods or passengers, rendez-vous between mobile agents, linkage of events with specific geographic information concerning places and their role within a logistic process, and so on.

Moreover, agent mobility must be coupled with data mobility. The next generation of sensor-networked infrastructures for ubiquitous computing shall support enhanced capabilities of sensing the mobile objects, thus allowing the collection of mobility data of higher precision and richness, covering not only positioning of mobile devices with reduced error and uncertainty, but also many further properties of mobile behaviours, recorded thanks to the enhanced capabilities of interaction with the surrounding environment.

The change in time and space of data distributions makes useless any algorithm based on static description of i.i.d. (independent and identically distributed) data. Moreover, this very change does not any more allow relying on a good asymptotic behaviour, as transitories become the norm.

As the task of knowledge discovery will result from the collective work of great number of agents and interactions, methods and approaches from the *complex systems* theory, *ant colonies*, or *swarm intelligence* may be mutuated as basic tools. All these methodologies must be investigated in depth, in order to understand if, how, and why they might help.

Another crucial issue to be resolved is the limitation in processing power, communication capability and data storage of small devices.

As the use of such devices is mandatory for the practical success of many applications, this issue appears to be the main bottleneck of future developments.

Finally, new interfaces between people and systems must be designed, based on multi-modality, ergonomic considerations, and cognitive models.

8   Overview of the Book

Part I of the book aims to provide a conceptual foundation for the new field of ubiquitous knowledge discovery, discussing the state of the art, highlighting challenges and problems, and proposing future directions. Although at some points technical examples are given for illustration, the aim of this chapter is non-technical.

While Part I is divided into individually authored chapters, it should be seen as a collaborative effort by the working groups of the KDubiq coordination action.[3] Hence, the chapters should be seen as an integrated whole. Part I of the book is structured as follows.

---

[3] In some cases, project partners provided input for some sections, but not for the whole chapter. Where this is the case, it is stated in the footnotes of individual sections.

Chapter 2 provides an overview on three distributed infrastructures for ubiquitous computing: (1) peer-to-peer systems, (2) grids, (3) sensor networks. The main characteristics of these infrastructures relevant for KDubiq are described. For each type of system, examples for distributed learning are given. These examples demonstrate that the choice of a platform has unique constraints for the learning algorithms, e.g., while for sensor networks energy consumption becomes a highly important criterion for algorithm design, in peer-to-peer systems it is the decentralized and non-reliable flow of communication. The examples show that in a distributed environment these considerations require new algorithmic solutions even for well-known tasks (e.g., eigenvector calculation), solutions that have no analogue in a non-distributed environment.

Chapter 3 discusses how the learning setting itself changes in an ubiquitous environment. A main observation is that in an ubiquitous environment resource constraints emerge that are quantitatively or qualitatively different from traditional data mining. E.g., sensor networks are limited several ways: compared to a desktop they have some orders of magnitude less computing power, main memory and storage; simultaneously, they are confronted with very large amounts of information to process. To handle this, a streaming setting is typically assumed. In this setting, aspects become central that so far exists more at the periphery of machine learning research: non-stationarity, change and novelty detection, incremental algorithms. Pushing this thought further, it can be seen that some of these scenarios are even outside the scope of current paradigms of theoretical learning theory (PAC learning, Statistical Learning Theory).

Chapter 4 defines general characteristics of data in ubiquitous environments. In all these scenarios the combination or fusion of different data sources is centrally important. KDubiq encompasses both Semantic Web issues and sensor applications. The chapter lays a foundation for analyzing both sources in a common conceptual framework. One important result is these seemingly very different data sources share many features. The discussion paves the way for analyzing hybrid systems where sensor data and semantic web data are combined.

Chapter 5 takes up the issues of privacy and security, arguing that they are critical for the deployment and user acceptance of KDubiq systems. After discussing basic concepts of security and privacy, the state of the art in security and privacy for ubiquitous knowledge discovery systems is discussed, focusing especially on the prospects of privacy-preserving data mining. Privacy-preserving data mining in a distributed streaming context is still in its infancy, and the chapter closes with pointing out directions for future research.

Chapter 6 is devoted to the human-centric view of ubiquitous knowledge discovery systems. In particular it discusses the impact of ubiquity in gathering requirements. In fact, in a KDubiq setting people play a much more direct role as consumers of this technology than in a traditional data mining setting. Moreover, this holds for users in the traditional sense, but it also calls for a reconsideration of the very notion of users. The chapter discusses the stages of requirement analysis, data collection, information presentation and interaction, processing, and evaluation. The chapter is necessarily somewhat different from the preceding ones because it does not deal with a specific technological concept like data or algorithms. (A description of the specific technological concept interfaces would be more fitting to a ubiquitous computing HCI text.) It is different because first, it takes a synoptic view in investigating the impact of the features of KDubiq (see Section 2), which are the subjects of Chapters 2–5, on the KD process from a people-centric perspectives. Second, to do so it needs to refer to methods and results from the interdisciplinary areas of HCI and Cognitive Modelling.

Finally, Chapter **7** contains a collection of potential application areas for KDubiq, providing pointers to the state of the art, to existing applications (if available) and to challenges for future research.

Part II contains selected approaches to ubiquitous knowledge discovery and treats specific aspects in detail. The contributions have been carefully selected to provide illustrations and in-depth discussions for some of the major findings of Part I.

The contribution by **Antoine Cornéujols** takes up in greater detail two fundamental challenges for learning in ubqiquitous environments, incrementality and non-stationarity, as described in Chapter 3. He surveys and systematizes the state of the art and suggests future directions and extensions, arguing that fundamentally new theoretical approaches for understanding the learning process and for designing new algorithms are needed.

The chapter by **Severo** and **Gama** is located in the same general problem space, but in a specific setting and with a specific algorithmic contribution. The authors investigate change detection in temporal data, when the assumption of stationarity is not met. This is one of most fundamental challenges for ubiquitous knowledge discovery, setting it apart from most traditional lines of research in data mining and machine learning (cf. Chapter 3). They propose a wrapper for a regression method which uses the Kalman filter and a CUSUM (cumulative sum of recursive residuals) change detector. The proposed method is very general and can be applied or extended in many specific contexts.

A third contribution to this topic, but again from a different perspective, is the paper by **Sharfman**, **Schuster**, and **Keren**. Their topic is to monitor changes in highly distributed data stream systems using a *geometric approach*. The basic idea is to decompose a global monitoring task into the monitoring of local constraints that can be checked at each node, without communication. This results in very good scalability properties. The methodological key challenge is that in this setting there is a fundamental difference between the monitoring of linear and non-linear functions. Previous work provides only ad-hoc solutions for the difficult non-linear case or, in most cases, does not address it all. The paper describes a general approach to this problem based on a geometric framework.

The three contributions nicely complement each other by discussing a general learning framework, and by adressing the temporal and the geometric dimensions of the problem space.

The contribution by **Inan** and **Saygin** addresses another highly important dimension of ubiquitous knowledge discovery: privacy. The general problem setting has been addressed in Chapter 5. In this chapter the authors investigate the problem of clustering horizontally partioned spatio-temporal data (the data sets at the nodes have the same attributes, but different instances) in a privacy-preserving manner. The approach is based on a protocol for secure multi-party computation of trajectory distances, serving as a building block for clustering.

The chapter by **Katharina Morik** describes a peer-to-peer Web 2.0 application for collaborative structuring of multimedia collections. As Chapter 4 explains, both data from mobile applications and from the web are important data sources for ubiquitous knowledge discovery. While the last chapter focussed on spatio-temporal data as they arise in mobile applications, e.g. from GPS sensors, this chapter complements the picture by focusing on web 2.0 data. It discusses the challenges coming from this data source and especially addresses the problem of machine learning in collaborative ubiquitous environments. A unique feature of this chapter is that the learning algorithms take account of the subjective nature of the collaborative learning problem.

The chapter by **Rasmus Pedersen** broadens the discussion and provides an overview on the topic of learning in micro-information systems. These are embedded information systems with contextual constraints in size, form, and functions. The chapter discusses the role and characteristics of ubiquitous knowledge discovery for these systems. It specifically discusses the Lego Mindstorms platform as an educational platform, and gives an example of a support vector machine implementation based on TinyOS. This chapter also serves as an excellent practical illustration how different in flavour the KDubiq approach is from desktop based learning environments.

The final chapter by Hillol Kargupta and co-workers describes the MineFleet system, one of the few commercially available ubiquitous knowledge discovery systems. Its aim is vehicle performance monitoring for commercial fleets. The paper describes the overall architecture of the system, business needs, and shares experience from successful large-scale commercial deployments.

To sum up, Part I of the book aims to provide a conceptual foundation for the new field of ubiquitous knowledge discovery, discussing the state of the art and proposing future directions, while Part II gives detailed technical accounts for this approach. We hope that the reader will agree with us that ubiquitous knwoledge discovery in many way holds potential for radically changing the way machine learning and data mining is done today, and share our excitement about this new field of research.

References

1. Liao, L., Patterson, D.J., Fox, D., Kautz, H.: Learning and inferring transportation routines. Artif. Intell. 171(5-6), 311–331 (2007)
2. Kargupta, H., Bhargava, R., Liu, K., Powers, M., Blair, P., Bushra, S., Dull, J., Sarkar, K., Klein, M., Vasa, M., Handy, D.: Vedas: A mobile and distributed data stream mining system for real-time vehicle monitoring. In: Proceedings of the SIAM International Data Mining Conference, Orlando (2004)

3. May, M., Berendt, B., Cornéjols, A., Gama, J., Gianotti, F., Hotho, A., Malerba, D., Menesalvas, E., Morik, K., Pedersen, R., Saitta, L., Saygin, Y., Schuster, A., Vanhoof, K.: Research challenges in ubiquitous knowledge discovery. In: Kargupta, H., Han, J., Yu, P.S., Motwani, R., Kumar, V. (eds.) Next Generation of Data Mining, ch. 7, pp. 131–150. Chapman & Hall/CRC (2008)
4. Mierswa, I., Morik, K., Wurst, M.: Collaborative use of features in a distributed system for the organization of music collections. In: Shen, S., Cui, L. (eds.) Intelligent Music Information Systems: Tools and Methodologies, pp. 147–176. Idea Group Publishing, USA (2007)
5. Flasch, O., Kaspari, A., Morik, K., Wurst, M.: Aspect-based tagging for collaborative media organisation. In: Proceedings of the ECML/PKDD workshop on Ubiquitous Knowledge Discovery for Users (2006)
6. Wurst, M., Morik, K.: Distributed feature extraction in a p2p setting - a case study. In: Future Generation Computer Systems, Special Issue on Data Mining (2006)

Ubiquitous Technologies

Assaf Schuster[1] and Ran Wolff[2]

[1] Faculty of Computer Science,
Technion – Israel Institute of Technology,
Haifa, Israel
[2] Department of Management Information Systems,
Haifa University,
Haifa, Israel
assaf@cs.technion.ac.il, rwolff@mis.haifa.ac.il

From a technological point of view, ubiquitous computing is a specific case of distributed computing. However, while the textbook definition for a distributed system is (Tanenbaum and van Steen)

> "a collection of independent computers that appear to its users as a single coherent system."[1]

Ubiquitous systems challenge this definition in various ways. First, some of those systems, wireless sensor networks, are collections of computers, each of which makes no sense as an independent computer. Second, in systems that rely on self-organization – peer-to-peer networks for example – the user is often the system itself rather than a human being. Finally, ubiquitous systems are often complex and dynamic systems: coherence, though it may be desired, is rarely achieved.

The traditional definition of Tanenbaum and van Steen usually assumed a machine is an independent computer, that communication is LAN based, that users provide the data as input and expect the output to be strictly defined by the program, and finally, that control is centered in the hands of the user or of a single administrator.

Another major characteristic of ubiquitous systems is that they are, in many instances, information systems; focusing on the collection and storage of information, information processing, and dissemination of the resultant knowledge. Ubiquitous knowledge discovery technologies can thus be said to emerge from the ways in which ubiquitous systems contradict, to various degrees, the assumptions which shape other distributed knowledge discovery technologies.

The discussion of knowledge discovery and ubiquitous systems can benefit if the architecture or technology of the system is first analyzed along six main axes. These are:

1. Device: what are the storage and processing capabilities of each device?
2. Communication: how do the different machines communicate with each other?
3. Users: who is the user and how does the user interact with the system?
4. Control: who controls each of the aforementioned components?

M. May and L. Saitta (Eds.): Ubiquitous Knowledge Discovery, LNAI 6202, pp. 19–39, 2010.
© Springer-Verlag Berlin Heidelberg 2010

5. Data: what data are stored on the system? What are its dynamics and its organization?
6. Infrastructure: What knowledge discovery infrastructure exist for the system?

The rest of this section describes technologies which were adopted for several exemplary problems in ubiquitous knowledge discovery. All the problems described here can be addressed using a distributed data mining approach, and are different in perspective from those addressed by multi-agent systems, as described in Chapter 3. Furthermore, since our objective is to describe technologies which can be carried over from one system to another, this section leaves out many considerations which are respective to specific resources; these are described in detail in Chapter 6.

We begin by describing the architecture of three ubiquitous systems in terms of the six axes. Then, we briefly demonstrate how data mining can be carried in those systems, pointing to key issues. Finally, we analyze at greater length the main choices available for the designer of a ubiquitous knowledge discovery system. We conclude with some visionary remarks on possible future advances in ubiquitous technologies, which may better facilitate better knowledge discovery.

1   Peer-to-Peer Systems

The Gnutella file exchange system comprises home PCs running one of several implementations of the Gnutella protocol. Communication in Gnutella is performed in an overlay network, where peers are designated neighbor peers, with whom they communicate over the Internet. Communication can be considered reliable, unless a peer goes down. The topology of the Gnutella network, and thus its membership, are determined by a protocol. Thus, no person has direct control over who the members of the network are.

The users of the Gnutella network access it through their client software. Thus, they are a homogeneous group (no super-users, or restricted ones). The client allows a user to share files on his or her PC with the other users, to search for files that other users have shared, and to download them.

Users have control of the data and the bandwidth requirement of their own PCs, and can also connect or disconnect to the system at will. Control over the rest of the system is enforced by protocol. The development of that protocol follows a quasi-democratic procedure in which various developers offer newer implementations and users vote by preferring to use one client implementation or another.

Data in Gnutella consists of the shared files, and the control data collected by the system. An example of that control data could be the trace logs of which files were downloaded by and from which peer. Several methods have been proposed to mine this control data. However, before we describe them, we need to discuss the dynamics of this data.

Gnutella users typically download files larger than a megabyte. At Internet speeds, one can expect a download rate of no more than one new file per minute.

Every trace record will contain the details of the file, the address of the source, and perhaps an indication of the success of the download. This can be calculated, for example, according to whether the user has subsequently tried to download the same file from another source. This record cannot be larger than a few dozens of bytes. Thus, an educated guess would be to bound the information rate per peer by as little as a few bytes per second, and the rate for a one-million peer system by few megabytes per second.

Finally, no knowledge discovery infrastructure exists in any known peer-to-peer system. Essential services such as the ability to sample random data or compute averages are not part of the protocol package and must be implemented as an application layer. This is despite the significant work which has been done on algorithms for knowledge discovery in peer-to-peer systems. There can be several reasons for this lack of infrastructure: For one thing, peer-to-peer systems tend to be application specific. Thus their architects are usually not interested in generic services such as knowledge discovery. When specific problems that can be solved using knowledge discovery arise (e.g., trust-management, see below), other solutions were often found to be no less effective. Finally, privacy has become a major issue with peer-to-peer data; many peer-to-peer systems are in dispute with the digital rights law and no privacy model yet exists which is suitable for peer-to-peer scales and dynamics and accepted by the community.

An application of peer-to-peer networks to music mining is described in Chapter 1, Section 6.2.

2   Grid Systems

Grid systems manage and run user jobs on a collection of computers called pools. These pools contain two main types of machines: *execution machines*, also termed *resources*, and *submission machines*, with whom users interact. Every actual computer can serve as both an execution and a submission machine. However, in practice, the vast majority of computers usually function only as execution machines. Additionally, a pool contains a handful of other machines which provide additional system services to which they are usually dedicated.

Submission machines are the user interface of the grid system. Mostly, this interface allows users to submit a batch job (or a collection thereof) to the system, and then track its progress through queues and execution. A submission machine is in charge of locating machines where the job may be executed, performing the required transaction with this execution machine, tracing the status of the job while it executes and managing reruns if the job fails to complete. Eventually, after a job has completed, the submission machine retrieves the job's outputs and stores them for the user.

The bulk of machines in a grid system are execution machines. An execution machine can be a standard workstation, a server, or even a supercomputer. Usually, it will be of a rather updated model (most organizations consider it uneconomical to maintain old model machines). Execution machines are typically connected to one another and to the submission machines via a local area network. When grid systems are composed of machines in several remote location their organization is usually hierarchical, and an effort is made to retain locality for the sake of faster communication.

One service which is today gravely lacking in most grid systems is a rich communications layer. User jobs can generally only communicate with one another via files. There is no systematic way for concurrent jobs to communicate. There is, in fact, no way to guarantee that two jobs will even execute concurrently. This is not accidental. One of the main features of grid systems is the relatively low availability of every one of the resources. Although this is compensated for by the sheer amount of resources, and thus the system as a whole has high throughput, it is still very difficult to reserve a group of machines which will be available together and will be in proximity to each other. Usually, availability and proximity can only be implemented by dedicating a group of machines to the same user, or by collecting machines one by one and keeping them idle until enough have been collected. Both methods are highly wasteful in terms of system throughput and are thus unpopular.

Execution machines share their computational capabilities with the pool according to policies set by their administrator. For example, the administrator may decide that its machine should only receive jobs submitted by a specific group of users and only if the machine has been idle for more than an hour. The owner can place further limitations on jobs, such as that they will be Java jobs requesting no more that 100 megabytes of memory.

As both submission and execution machines pose requests and requirements, the task of matching jobs to machines requires a third component — the matchmaker. The matchmaker collects information about all the pool participants, calculates the satisfiability of their demands, and notifies the submission and the execution machines of potentially compatible partners.

An additional important grid service is the ability to plan and execute workflows – groups of jobs that depend on one another. Condor [2], for example, provides a service which supports synchronization (i.e., waiting for a job to terminate before starting another one) and limited control structures. In this way, for instance complex queries and even data mining algorithms can be implemented on top of a data grid.

Finally, in order to better utilize resources, a grid system has to efficiently schedule the jobs and balance the load among the execution machines. Load balancing and scheduling differ slightly in different grid systems. Condor, for example, retains a queue at every submission machine. Submission machines autonomously schedule jobs to the machines allocated to them by the matchmaker. Intel NetBatch, on the other hand, implements a hierarchical matchmaker.

Either way, grid systems are subject to more stringent control than the peer-to-peer systems described earlier. The several organizations or several parts of the same organization, which contribute machines to the system usually maintain a predictable level of cooperation or are bounded by service level agreements of one kind or another.

One important variant of grid systems is *data grids*. shared by an execution machine is not its CPU cycles but the data it stores. A job, in a data grid context, can be an SQL query intended for this data. Data grids are important in several contexts: when, for example, the data shared by resources is too large to be replicated elsewhere, or when ownership and privacy concerns prohibit replication of the data or parts of it, but do not prohibit providing some kinds of queries – like aggregates and small samples. In this case, the system can enforce limitations on queries which the sharer of the data deems acceptable.

The amount of data stored in data grid systems can be huge. The US National Virtual Observatory is projected to share petabytes of sky images (in various spectra) at a rate which will increase by as much as seven petabytes annually by the year 2012. Likewise, the Large Hadron Collider is projected to produce some eight petabytes annually beginning in 2008. Both projects rely on data grid technology for sharing and processing the data. Even computational grid systems create large amounts of data. For instance, it is estimated that the control data generated by the 1,000 machines in a Condor pool amounts to one hundred gigabytes a day – or about 30 terabytes annually.

Knowledge discovery is envisioned as one of the "killer-applications" of data grid systems, largely because the data stored on those systems can only be surveyed and analyzed by efficient software. Browsing it manually would be both infeasible (because of its size) and pointless (because it is beyond human capacity to generalize out of so much data). Additionally, many knowledge discovery procedures can be implemented as sequences of interdependent jobs, which is very suitable for grid systems. Furthermore, by pushing the processing of data to the place where it is stored one might be able to solve data ownership and privacy concerns which sometimes hinder the ability to share data on data grids.

Several projects (e.g., NASA's Information Power Grid [3], Discovery Net [4] and Grid Miner [5]) have surveyed the possibility of performing knowledge discovery on grid systems. Lately, a thorough effort to implement a knowledge discovery infrastructure atop grids was made in the EU DataMiningGrid project [6] [7]. The infrastructure includes services which discover both data and algorithm implementation resources, and tools for building and executing the complex workflows that implement a knowledge discovery process. Below we describe the application of an early version of this architecture to solving one grid management problem: the discovery of misconfigured grid resources.

2.1   Example - Misconfiguration Detection in Condor

Grid systems are notoriously difficult to manage. First, they suffer many faults, because of the huge heterogeneity in the hardware and the heterogeneity of the applications they execute. Faults also arise because grid systems often pool together resources that belong to several administrative domains, such that no single authority has the ability to enforce maintenance standards (e.g., with respect to software updates). Once problems occur, the huge complexity of the system makes it very hard to track them down and explain them. Maintenance, therefore, is often catastrophe driven and rarely preventive.

In GMS [8] knowledge discovery was suggested as a better means for detecting misconfigured machines. It is suggested that by analyzing system logs, state indicators (e.g., CPU and disk usage) and configuration variables, machines whose behavior is anomalous – outliers – can be detected. In itself, this "black-box" analysis method is not new and was tried on many software systems. What is new is the way GMS approaches data collection and implementation.

In previous work on distributed systems data collection posed heavy constraints on the kind of data analysis that is possible. For instance, in eBay [9], indicative data was streamed to several dedicated machines for analysis. The resultant bandwidth consumption was so large that the data had to be limited to just six attributes. This might have been sufficient because the eBay system is monolithic and the jobs very similar. However, it is hard to see how such an approach could work for a grid system in which every job might define dozens of attributes, and system configuration requires hundreds. Instead of centralizing data, GMS proposes to leave it where it originates – on the machines. Then, GMS proposes to implement the analysis distributively. Another angle of this approach is that the distributed analysis is conducted via a collection of recursive interdependent jobs (or workflows), which operate just like any other job in the grid system. Thus, analysis requires no special implementation and no dedicated resources.

3 Wireless Sensor Networks

Wireless sensor networks (WSNs) are collections of dozens or hundreds cheaply produced, easily deployed, sensing devices. They are intended for use in areas where it is not economical or not possible to install fixed power and communication infrastructures. The most obvious examples is hostile or hard-to-reach terrains, such as battlefields, natural disaster areas, mountainous regions nature, or the inside of an oil well. Since an infrastructure does not exist, WSN devices are battery operated and communicate wirelessly with one another or with a base station.

Battery power is the single most important resource of wireless sensors (also known as motes.) Power consumption places stringent limits on the mote's lifetime and is a key determinant of the cost effectiveness of the system. When the system dissipates a large portion of its combined energy, motes will begin dying out, a process which will at first limit coverage and eventually render the system dysfunctional. The nature of the applications considered does not permit manual replacement of batteries. Furthermore, in most cases it is impractical to complement a mote with a renewable source of energy (e.g., solar panels). Thus, the only way to avoid system breakdown when battery power dissipates is to deploy additional sensors.

Mote hardware is geared towards energy conservation. Most often it is combined of an energy efficient 8-bit processor, a near-range transmitter / receiver, and a small (128K) flash memory bank. This basic package is then augmented with a sensor board which may include sensors of different types: light, temperature, PH, barometric pressure,acceleration/seismic, acoustic or magnetic. This hardware greatly limits the kind of computations which can be performed on a mote. For instance, [10] shows that Fast-Fourier Transform (FFT) of just 4,000 measurements may take more than 30 seconds and is thus impractical.

Communication in wireless sensor networks relies on the short range radio transceivers of the motes. Motes must be deployed densely enough to permit every one of them to exchange information with at least one, and preferably a few, other motes. For a mote to send data to motes out of its communication range, or to a remote base station, other motes need to serve as relays. The topology of the network is in general unknown to the sensors. Furthermore, it is expected to change as sensors die out, or as the environmental conditions (i.e., RF noise levels, rain, etc.) change.

Communication in wireless sensor networks is more costly than most other operations. Not only does it increase current draw by an order of magnitude, it also requires neighboring motes to invest valuable energy in receiving, processing, and sometimes forwarding the message. System design, therefore, tends to focus on reducing the number of transmissions. This focus is evident both in the planning of communication services and in the implementation of data processing algorithms.

In terms of communication infrastructure, most wireless sensor networks prefer not paying the overhead required to establish reliable communication. For many applications, a missing subset of the messages is acceptable, but if not, the cost of ensuring message reliability is unacceptable. Moreover, many networks avoid the routing protocols required to ensure message addressability: Rather than supporting the ability of motes to address a message to any other mote, they make do with addressing messages to either all motes (broadcast) or just to the base station (convergecast.)

The communication pattern of data processing algorithms has changed focus in recent years. Originally, wireless sensor networks followed the simplistic approach of centralizing all data to a base station (possibly after in-mote filtering and summarization; see [11].) This approach positions the designer before a vicious dilemma: if much of the data is centralized, then message forwarding becomes the main energy consumer of the network and its cost-effectiveness is hampered. If, however, excessive in-mote filtering and summarization are used, then the results fall short of the accuracy obtained by ideal computation over fully centralized data. To avoid this dilemma, data processing algorithm designers have shifted to an in-network approach. The idea of this approach is to compute exact ideal results or good approximations thereof by pushing functionality down from the base station and into the network itself. In the following section we describe several algorithms for such in-network data processing.

Unlike the systems described in previous sections, wireless sensor networks have a single user entity. This is mostly because the high cost of each query to the system compels meticulously planning and monitoring by the system administrators. The administration of each system is handled by a single authority. The only exception considered to the single-user / single-administrator control structure is in military applications, where it is sometimes assumed that an adversary could seize control of one or several motes, and manipulate them for his own goals – thus creating two conflicting user and administration domains. Still, this exception to the conventional model has, so far as we know, only been tested theoretically.

The sole purpose of deploying a sensor network is to collect environmental data. The rate at which this data is sampled is determined by the application. For instance, a sensor network for detecting changes in thermal conditions can afford a rather slow sample rate, whereas one intended for infra-sound signals (e.g., in volcano monitoring applications) may need to operate continuously. It is almost never feasible to channel all the data to a base station. Thus, as mentioned above, wireless sensor networks rely on either summarization or in-network computation to reduce the system-level data rate far below the sum of single mote level data rates. In most systems, every mote collects the same set of measurements – that is, the data has the same scheme in every mote. However, most data have strong spatio-temporal affinity – the data of nearby motes collected at about the same time is more closely correlated than that of distant motes collected over different time periods.

Implementations of knowledge discovery services for wireless sensor networks are few. There is ample work on analysis of sensor data using various knowledge discovery methods. However, this work, by and large, has taken one of two tracks: on-sensor analysis of locally sampled data or central analysis of the entire network's data after its collection. Neither track amounts to what we would consider knowledge discovery infrastructure. For this, one would need tools for the in-network analysis of data collected by multiple sensors. Such tools can rely on existent work on database infrastructure for wireless sensor networks [19] and on work on in-network computation of knowledge discovery methods such as those described below.

3.1   Example - Source Localization in Wireless Sensor Network

Source localization is the problem of computing the coordinates of the source of a signal from measurements taken by multiple motes distributed in its vicinity. It is a central problem for wireless sensor networks because it is tightly related to intrusion detection – one of the major applications of wireless sensor networks. Source localization methods differ first and foremost by the mechanism they employ to compute the location. One approach relies on the fact that a signal emitted by the source diminishes proportionally to the distance according to known formulas which depend on the type of signal (acoustic, electro-magnetic, etc.) and the medium (air, water, ground, etc.).

In an unrealistic noise-free environment, exact measurement of the signal strength from three sensors can lead, by triangulation, to the exact source (in the plane) of the signal: Each sensor can draw the surrounding circle on which the source might lie, and the point of intersection of the three circles denotes where the source is. An extra measurement would be needed if the source amplitude is unknown. Likewise, localization in space would require an additional measurement. In a more realistic scenario, one which still assumes a single source but permits random noise, maximum likelihood methods can be used: the likely source would be the point in space minimizing the error attributed to the noise portion of the signal.

The problem can be set as one of quadratic minimization, which has two properties that make it simple to solve in a distributed environment: The first is that it only has a single local minima – the optimal solution – and, thus, can be solved using gradient descent. The second is that the error function is decomposable; thus, gradient descent can be carried out in increments computed independently by individual motes; each of whom only uses the current assumed location and its own data.

In [12], an entirely decentralized algorithm is presented: it takes advantage of these two properties. As such, it is likely to be superior to any centralization based algorithm, at least if the requirement for localization error ϵ is moderate, which is often the case (especially when one takes into account the existence of additional sources of noise). Furthermore the algorithm requires only local, and not global addressability. It should be further noted that, when building from local ordering (of a mote respective to its neighbors), global enumeration is not hard to compute. Still, the algorithm does have several drawbacks. Most notably, it suffices for one message to be lost, or one mote to consume all of its energy, for the algorithm to terminate prematurely. Furthermore, the initial choice of received signal strength as the input of the algorithm is problematic because it requires hardware that is costly and more sensitive than that required by other mechanisms (such as time difference of arrival – TDOA).

Table 1. Summary comparison of main ubiquitous technologies

	Peer-to-Peer Networks	Grid Systems	Wireless Sensor Networks
Device	Home PC	From PC to supercomputers	Low power, limited memory
Communication	Internet-like	File exchange via centralized services	Radio based, sparse and costly
Users	Multiple consumers	Multiple engineers or scientists	Single engineer or scientist
Control	Each user controls own device	Multiple administrators bounded by SLAs	Single administrator
Data	Typically multimedia, slowly changing, same schema to all	Typically scientific multiple schema	Environmental measurements highly dynamic, same schema to all
Infrastructure	None	Knowledge discovery web services	Database services, no knowledge discovery support

4   Design Choices

In the following section we reiterate the review of ubiquitous technologies. However, instead of focusing on the different systems, we focus on design issues that repeat themselves in many, if not all, of those systems.

4.1   Device Heterogeneity

The devices used in ubiquitous system are never really uniform. The reasons for that are four: One, ubiquitous systems are long lived, and in a long lived system the history of various devices affects them. Two, ubiquitous systems are (in general) large, and thus tend not to be based on devices with equal hardware specification. Three, ubiquitous systems are, more often than not, composed of volunteered resources, actually owned by a variety of entities; hence, it is hard to enforce standards on the device software and service level. Last, ubiquitous systems span large geographic and topological domains. Thus, location and relative distances play an important role in the performance of each device.

Examples for this heterogeneity persist through all of the systems presented here: Past activity is a central determining factor for the power reserves of wireless sensor motes, and for the reputation of peers in a peer-to-peer network. Heterogeneity of the device hardware is imminent in peer-to-peer networks and in grid systems. Variation of service levels is actually a main feature of grid systems and of most peer-to-peer file-sharing networks, and most of these systems are multi-platform (i.e., Linux, Mac, and Windows.) Topology and relative distances play a key role in multi-pool grid systems and in wireless sensor networks; and take center stage in mobile ad-hoc networks.

The main effect of device heterogeneity on ubiquitous systems is that it renders the simplistic uniform architecture inefficient. A wireless sensor network which assigns the same role to a sensor regardless if it has two months or two hours of operation left, and regardless if it is near or far from the base station would seize to function – as a system – much earlier than one which does take into account these variations. A grid system which takes no notice of resource hardware and software specifications in job assignment is so degraded that it can only be theoretical. Even peer-to-peer networks, where the trend towards uniform design has been the strongest, see great improvement in performance when peers' roles are varied respective to their bandwidth or reliability.

Ubiquitous systems have adopted various approaches to take advantage of device heterogeneity. Mainly, the approach is to place different and more central roles to exceptionally well equipped devices. This solution is the basis for the assignment of super-peer roles to reliable and well-connected peers in peer-to-peer networks, for the assignment of sentries in many wireless sensor network applications, and for the designation of reliable and highly available devices for match-maker and collector roles in grid systems. Whatever their name is, these designated devices execute a different, usually more costly, protocol than the majority of devices. Often, choosing which devices to designate becomes an important system level activity in its own right.

4.2   Communication Limitations (Reliability, Timeliness, and Cost)

To see how limited communication in ubiquitous system is, one only has to review the famous 8 fallacies of distributed computing (Peter Deutsch, James Gosling [13]):

1. The network is reliable.
2. Latency is zero.
3. Bandwidth is infinite.
4. The network is secure.
5. Topology doesn't change.
6. There is one administrator.
7. Transport cost is zero.
8. The network is homogeneous.

This list is often quoted as a warning to architects of networked systems. The special characteristics of ubiquitous systems highlight each and every one of the topics on the list (the last one has been discussed above).

Network reliability comes at a cost which often is too high in ubiquitous systems. Grid systems, for example, often use unreliable communication between resources and the resource broker because (1) it is impossible for the resource broker to manage thousands of TCP connections, and (2) there are so many resources that occasionally missing one because a message is lost is unimportant. Wireless sensor networks, too, rarely use reliable communication. This is because the energetic costs of acknowledgment of messages which got to their destination and retransmission of those which have not are far too high.

The latency of sending a message in a ubiquitous system can be very significant. In distributed hash table based peer-to-peer, for example, the promise is to deliver a message in $O\,(logN)$ hops, where every hop involves sending a message to a random peer. Assuming a million peers, this means that sending a message to a peer would translate into exchanging 20 equivalent Internet messages between the peers which route the message. I.e., a latency at least 20 times larger, on average, than one would expect.

Bandwidth is another pressing matter for ubiquitous systems. It is one of the main obstacles to scaling up wireless sensor networks – because with scale the number of message collisions increases to the point where communication becomes nearly infeasible. It is also the showstopper preventing global search in a Gnutella-like peer to peer system. Consider a peer connected to an Ethernet network and dedicating 1 mega bit per second to the system. With this budget it can accept up to 666 1,500 bit long Ethernet messages. Now, if the system supports ten million peers, and if each peer would perform global searches such that every such search requires sending just one message to every peer, then the number of searches a peer could make would be limited to less than six daily.

Security in ubiquitous systems is a grave problem with just few solutions and worthy of a long discussion (see Chapter 5). The core of the problem is that because of their voluntary and dynamic nature, most ubiquitous system are open. Anyone can add resources to a system with just little difficulty. This has been widely taken advantage of in the context of file-sharing peer-to-peer networks, where it is known that the music industry both infiltrated networks to gather information about right violators, and "poisoned" the system with fake files containing degraded or malicious content. Wireless sensor networks, also, face grave security risk because of their use in hostile terrain. It is possible that the enemy may seize control of a mote and either manipulate its input or otherwise harm the system. Last, grid systems, like any other Internetworked system, have been breached in the past. Most notably, the PlanetLab global testing environment is known to have been breached [14] more than once. Such breach can, theoretically, permit the attacker to use the entire resources of the grid system to launch attacks of unprecedented scale.

The topology of any ubiquitous system is extremely dynamic. Resources constantly fail. Machines that recover (if they ever do) may have different connectivity and sometimes a different identity altogether. Routing usually has to be ad-hoc, because the value of a precomputed routing scheme degrades fast. In many systems, e.g., in Gnutella, routing is abandoned and the system relies on flooding instead.

The number of administrators in a ubiquitous system varies. In wireless sensor networks, there is usually just one body of administrators. In grid systems, there are many, but they are correlated and tied by service level agreements. In peer-to-peer systems, administration is often limited to tuning the underlying protocols which dictate, for instance, the load balancing between peers. However, since the system is open, anyone implementing the protocol is in-fact an administrator for those users who choose to use its client. Thus, not only the number of administrators is large, they are not always known.

Last, the cost of transport is not zero. However, in ubiquitous systems, that cost has usually already been paid for. I.e., Skype relies on the user to already own an Internet connection anyhow, and the transceiver of a wireless sensor network is part of the mote – or there would be no network to talk of. Occasionally, a system may put so much load on the network that it would actually require its upgrade. This is true, for example, with file-sharing peer-to-peer networks which have become one of the major users of bandwidth.

4.3   Centralization and De-centralization

There is an inherent tension, in nearly any distributed system, between forces motivating centralized architecture and those motivating de-centralized architecture. In ubiquitous systems, centralization can be prohibitively costly. Furthermore, centralization is often motivated by the wish to achieve some kind of a global agreement, synchronization, or coherence. Since ubiquitous systems change rather rapidly, these are almost certainly impossible and hence this motivation is missing. Another motivation is the desire to use every resource available, and hence the need to keep stock of resources and their state. However, ubiquitous system often contain so much redundancy (e.g., files in a peer-to-peer system, machines in grid systems, etc.) that this motivation too becomes weaker.

For all those reasons, ubiquitous systems often adopt de-centralized solutions and are willing to easily sacrifice unreachable goals such as globally accurate results.

One of the simplest approaches for de-centralization is building an hierarchical system or a hybrid one. In a hierarchical system, e.g., in the EGEE or Netbatch grid systems, there is a strict control structure in which resources are divided to pools, each controlled by its own submission and job-scheduling machines. Those controlling machines are in turn centrally or hierarchically controlled. In a hybrid system (more typical with peer-to-peer or wireless sensor networks), there are still low level and high level machines. However, the association of low level machines to high level ones is flexible and may dynamically change. Furthermore, the high level machines themselves are not necessarily centrally or hierarchically controlled but rather often form their own network. This is, for instance, the structure of peer-to-peer networks who employ super-peers, or of wireless sensor networks which integrate high powered relays, connected to the base station.

The main disadvantage of hierarchy, like any other control structure, is that it is prune to failure, or at least wasteful to maintain, when the dynamics of the system itself are intense. On such conditions approaches which use flooding, or a more controlled variation of flooding like gossiping and local-thresholding are often best. Flooding is used, for instance, for searching in some peer-to-peer networks; gossiping for mobile ad-hoc networks; and local-thresholding for wireless sensor networks and grid systems.

One last approach for decentralization which is widely practiced in peer-to-peer networks is distributed hashing. In distributed hashing, which is the basis for such networks as Chord, the system is highly structured, yet non-hierarchic. Distributed hashing therefore requires a lot of maintenance, and may be as costly as centralized approaches, however it is by far more resilient to partial failure.

4.4 System and Data Dynamics

Small-scale distributed systems, built of industry-standard hardware and software, have downtime which measures in small fractions of a percent. Furthermore, with the tight control characterizing them, fail-over mechanisms are often introduced which decrease downtime even further. Therefore, knowledge discovery research tends to overlook the issue of failure altogether.

In ubiquitous systems, downtime of components is an integral and unavoidable part of system's behavior. For instance, if every user of a peer-to-peer system goes down for two minutes a day (e.g., resets the network connection or reboots) then more than a thousand of each million peers are down on every instant. Similar examples can be given for wireless sensor networks (where it is expected sensors will gradually die out) and for grid systems (where the owner of a resource regularly takes precedence and evicts all other users.)

The presence of downtime in ubiquitous systems affects knowledge discovery in two ways: One is that occasionally parts of the computation abruptly terminate. Furthermore, such terminations are so frequent in most systems, that fail-over becomes impracticable. The second way is that parts of the data – the input – becomes inaccessible. This poses the question what should the output be when just a selection of the input is accessible, and how can accuracy and coherence be guaranteed when this selection changes constantly. The dynamics of data change are often further complicated when even the data in those parts of the system which remain on and connected changes during calculation of the output.

Hard as these problems may seem, they are not new. It does well to re-member that similar problems characterize the Internet infrastructure: Routers, too, are abundant and suffer a high failure rate, with failures, too, being possi-bly correlated. Routing data is, too, distributed and dynamic; and the functions computed by the Internet infrastructure are not conceptually that different from those computed by data mining algorithms: Selecting the next routing hop which will minimize the overall cost of routing a packet is not without resemblance to selecting the next feature by which to split a set of learning examples such that the gain would be maximal.

Reviewing the concepts developed in Internet routing protocols can be highly educational, as those protocols operate, for more than a decade now, in scales of millions of routers, and with data and topology that are no less dynamic than those of a peer-to-peer network. Routing protocols usually do not aim for optimal routing – in a dynamic system the optimum can rarely be computed. Instead they are concerned with the robustness of the solution: how bad might it turn out to be with respect to reasonable scenarios, what are these worst scenarios, and how can those be avoided.

The concept of robustness is not new for the knowledge discovery field. Many of the principles developed elsewhere can and should be imported to ubiquitous knowledge discovery: Avoiding outliers, for example, can be central for elimi-nating the adverse effects of failures. Relying on ordinal, rather than cardinal, statistics can greatly increase the stability of the solution when faced with rapid, yet stationary, change in the data. Finally, using random sampling as a way to avoid bias can undo the problems caused by of the spatio-temporal affinity of data, discussed below.

4.5   Spatio-temporal Data Affinity

The data collected and stored by ubiquitous systems often depends on the lo-cation and time of collection. This is true for wireless sensor networks, where data points sampled nearby and at approximately the same time are usually more correlated than data points sampled far from each other and on different times; it is also true for file-sharing peer-to-peer networks where, for example, a musical trends often prevails at a certain region and time – dictating which files are shared and what queries are posed to the system; grid systems, too, often have data which presents spatio-temporal patterns because of the joint usage of local resources (a telescope shared by all nearby universities, for example) or standards (e.g., Metric vs. Imperial units.)

The spatio-temporal affinity of data has two opposite effects on knowledge discovery: Applications which search for global knowledge should be careful not to ignore this bias. Extracting a uniform sample from data spread on a peer-to-peer system, for instance, can become a real challenge [15]. For instance, applications which, unwisely, relate to data collected in few nearby sites as a representative sample of the global data might be gravely misled. On the other hand, the spatio-temporal affinity of data means that all data points relating to one another are collected near one another. This can extremely expedite applications which search for outliers, for example, or are interested in clustering similar points.

It is important to note, in the context of spatio-temporal data affinity, that the basic design of some systems (namely, distributed hash table based peer-to-peer systems like Chord), seems to contradict any effort to make use of locale. This is because those systems create an overlay, and a data placement scheme, which distributes data to places independent of its origination point and connects peers to one another at random. Justified as this design might be, it seems mostly harmful in terms of data analysis. The topic of autocorrelation and non-stationarity is further discussed in Chapter 3.

5   Putting Things into Context – Mining Association Rules in Peer-to-Peer Networks[1]

This section discusses an exemplary knowledge discovery application, that of finding association rules, in the context of a representative ubiquitous system – a peer-to-peer network. The purpose of the discussion is to familiarize the reader with the technological challenges that are often raised by such applications and with the architectural and algorithmic tools that have been developed to address them.

Consider a file sharing network such as the Gnutella network, eMule, or Bit-Torrent. Those networks (mentioned earlier in Chapter 2) often span millions of peers. Each peer participates in the download, upload, and storage of thousands of files. In the process of doing so, the network accumulates valuable data: which files are often downloaded at the same day (e.g., consecutive parts of a series), which are the most popular, etc.

Some of this data can be analyzed for valuable patterns: If the system learns that a peer which downloaded file A would soon download file B it could pre-fetch file B and by that reduce wait time. Alternatively, the system can produce recommendations to its user to download one file because "users of your profile tend to". Thereby the user experience of the system would improve and it would gain popularity – a critical factor in peer-to-peer systems.

Such processing relies on commutative processing of user actions. To produce the association rule "users which downloaded file A tend to download file B as well" statistics must be gathered on the global trends of users. In terms of classical knowledge discovery the files downloaded by each user can be looked at

---

[1] This section is authored by Ran Wolff, Department of Management Information Systems, Haifa University, Israel.

as a transaction (similar to the content of a market basket) and the rule needs to be both popular with many users (i.e., with frequency above $minFreq$) and statistically valid (i.e., with confidence above $minConf$).

The main difference between the classical formulation of the problem is in the definition of the dataset. In the classical case the group of transactions is known in advance before the algorithm is executed. In the peer-to-peer case, the dataset is defined by the group of peers. This group is both dynamically changing – peers come and leave at a rate of hundreds per second – and never actually known to any single peer. To add to the ambiguity, the data at each peer can be thought of as constantly changing because peers keep on downloading files during knowledge discovery. The result of association rule mining in a peer-to-peer network is thus ad hoc and constantly changing.

To add to the difficulty of defining the requested functionality of association rule mining one should also consider the significant technological difficulties. The huge number of peers and frequency of failures (peers leaving the system) dictates that synchronization is impossible. The scale of the system also dictates that broadcast messages need be rare or at all none existent. A comparison to work on small scale distributed association rule mining would reveal that both synchronization and broadcast are heavily relied on in those settings.

5.1   A Peer-to-Peer Association Rule Mining Algorithm

The Majority-Rule algorithm [16] presented next addresses these difficulties in three conceptual steps. The first step is to modify the problem slightly so that it can be solved in a peer-to-peer setup. Noticing that an exact set of association rules can only be computed if the data stops changing, which never occurs, Majority-Rule permits every peer to compute a different set of rules. That set of rules may have some rules missing in it, or may have rules in it that are not present in the data. What Majority-Rule does require, however, is that whenever the data stops changing for a period the solution of every peer would converge to the correct one – the same solution that a conventional association rule mining algorithm would compute if given the entire data. This property is termed *eventual correctness*.

The second conceptual step is to reduce the association rule mining problem to a large collection of distributed majority votes. Taking for example a rule of the form $A \Rightarrow B$ which means that events A and B tend to co-occur more frequently than $minFreq$ and that furthermore when the event A occurs B also occurs at frequency greater than $minConf$ Majority-Rule suggest that the rule is tested with two votes: One in which peers vote whether A and B co-occurred and a majority of at least $minFreq$ is demanded; the other in which the peers for whom A occurred vote on whether B occurred too and a majority of $minConf$ is required. Clearly, if the required majority is achieved in both votes the rule $A \Rightarrow B$ holds and if in even one of the votes the majority is not achieved then the rule does not hold.

It is left to determine which votes to hold. Majority-Rule solves this by directing every peer to initiate new votes based on its current ad hoc solution. I.e., if at a given moment a peer computes that the rules $A \Rightarrow B$ and $A \Rightarrow C$ hold then it would initiate a vote on the rule $A \Rightarrow B \wedge C$. Note that at a given moment some peers may dispute that $A \Rightarrow B$ hold and thus this initiation process requires a certain amount of coordination between peers. However, because of the properties of the majority voting algorithm described next, this coordination requires no synchronization or broadcast.

The third conceptual step in Majority-Rule is the description of a *local* majority voting algorithm. Local algorithms are a breed of extremely efficient distributed algorithms. The distinctive note of all local algorithms is that each of them is based on a local stopping rule: A condition which can be evaluated independently at each peer based on the data it has and which, when it holds, guarantees a global property. The existence of the local stopping rule gives rise to simple algorithms in which every peer tries to stop based on its current data. If it fails, it sends its data to its neighbors and waits for updates to arrive. If it succeeds, then it does nothing unless one of its neighbors alters its data. It is easy to show that those simple algorithms always terminate with in one of two situations: One is that all peers received data which permits that they stop – in which case the local stopping rule dictate that the global property holds. The other is that all peers shares the entire data – in which case every peer can independently compute the property.

Perhaps the simplest example for a local algorithm is one in which every peer has an input of either zero or one and they all need to compute a logical Or over all inputs. The algorithm is to compute an Or over your input and every input received from any other peer and then send the result once to your neighbors if it is one. Clearly, if all inputs are zero then no communication takes place and the correct result is computed at each peer. Similarly, if all of the peers have an input one then a single round of communication would take place after which all peers would compute the correct answer. Notice that if an insignificant minority of the peers have input zero then the number of communication rounds would not increase by much.

5.2   A Local Majority Voting Algorithm

The local majority voting algorithm is just slightly more complicated than the logical Or algorithm. In local majority voting too, every peer has an input of one or zero. Peers receive and send collections of votes to their neighbors. Every peer tracks the number of *excess votes* they have. For peer i, this number is denoted Δ_i. The number of excess votes is computed as the number of one votes the peer has divided by the majority threshold minus the total number of votes the peer has. For instance, in a majority vote in which the majority threshold is 50% a peer i whose vote is one and who received three one votes and two zero votes from its neighbors has $\Delta_i = (1+3)/50\% - 6 = 2$. In another vote, in which a majority of 33% is needed, peer j whose vote is one and who received four zero votes has $\Delta_i = 1/33\% - 4 = -1$ because one additional vote would make the vote tied. In addition to the total number of excess votes, the peer also keeps track on the number of excess votes it has sent to every one of its neighbors and the number of excess votes it had received. A peer i which reported its neighbor j that it has two excess votes and was reported that j has minus three excess votes has $\Delta_{i,j} = -1$.

The algorithm is event based. Whenever peer i's Δ_i changes it inspects the position of the new Δ_i vis-a-vis all $\Delta_{i,j}$ the stopping rule dictates that it is okay not to update a neighbor j with the new votes that caused the change in Δ_i if and only if Δ_i is as extreme than $\Delta_{i,j}$. I.e., $\Delta_i \leq \Delta_{i,j} \leq 0$ or $\Delta_i \geq \Delta_{i,j} > 0$. otherwise, peer i would send to peer j all of the votes it received from peers other than j. An example of the response to one such event can bee seen in Figure 1.

Like the logical Or algorithm described above, the majority voting algorithm terminates on one of two cases: Either all peers have Δ_i which is more extreme than any $\Delta_{i,j}$ they have, or all peers have collected all votes. In the former case, it can be shown that the majority is of ones if $\Delta_i > 0$ and of zeros if $\Delta_i \leq 0$. In the latter, all peers got all votes and can calculate the result.

5.3   Limitations and Extensions

The majority voting algorithm described above has several limitations: Firstly, it requires that the network connectivity – the organization of neighbor peers – would be that of a tree. This may be trivial to guarantee in structured peer-to-peer systems (e.g., in DHT based systems such as BitTorrent) but is quite hard to guarantee in unstructured systems such as Gnutella. Further work has extended the local majority voting algorithm to general graphs, albeit at a significant communication and synchronization cost [17]. Another limitation is that the algorithm might fail if even one message gets lost. Severe as this limitation appears it can be done away simply by including acknowledgements between neighbors. The cost of that is negligible.

Also in further work, the same majority voting algorithm was used as a building block to many other data mining applications: facility location [18] and classification [19] to name just two. Even more importantly, the majority voting algorithm was generalized into an algorithm [20] which can test whether the average of input vectors in $\mathbb{R}^d$ resides inside any convex region in $\mathbb{R}^d$. This generalization served as a building block for other data mining applications such as Principle Component Analysis (PCA) and k-means.

Another interesting extension of this type of algorithms is towards semi-local algorithms. In semi-local algorithms, some data is broadcasted to all peers and serves to compute the result while the rest of the data is processed locally in order to validate the result. Semi-local algorithms were presented for outliers detection [21], k-means, and for the approximation of the mean vector of distributed data.

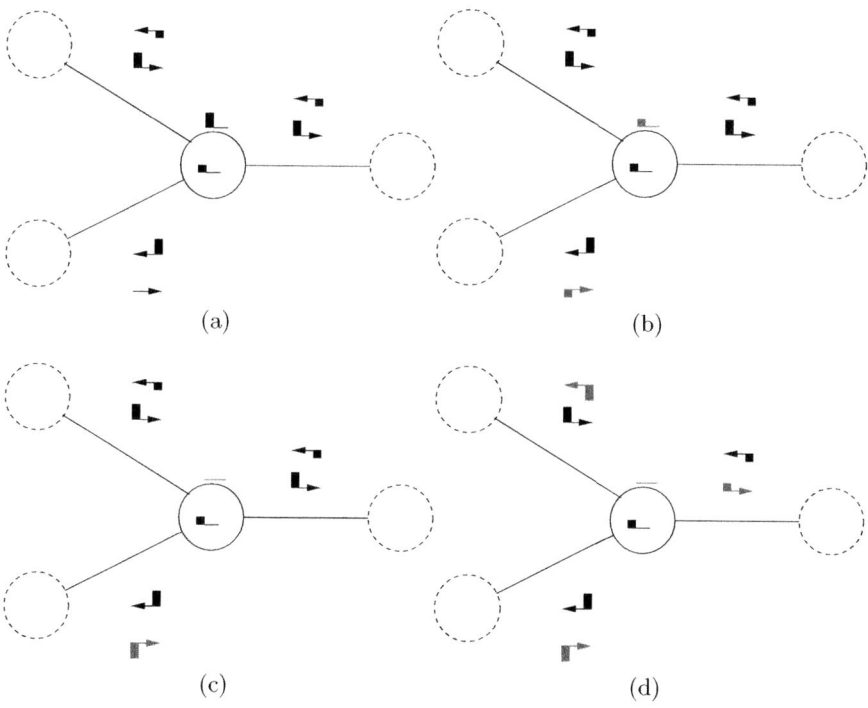

Fig. 1. A peer is described in steady state in Figure 1(a). Its own vote is one and it received an excess of -1 vote from its right neighbor (call it peer j), 0 excess votes from its bottom-left neighbor (peer k), and 2 excess votes from its top-left neighbor (peer ℓ). In total it has excess votes so $\Delta_i = 2$. Since it has sent two excess votes to its right neighbor (for a total of $\Delta_{i,j} = 1$), two to his bottom left ($\Delta_{i,k} = 2$) and -1 to the top-left ($\Delta_{i,\ell} = 1$). Thus, its current excess votes are as extreme as that of every one of its neighbors and it needs not send a message. Then, in Figure 1(b) an excess vote of -1 is received from peer k. This causes $\Delta_{i,k}$ to become 1, and Δ_i to become 1. Since Δ_i is still as extreme as that any of the neighbors this event causes no response. Next, in Figure 1(c) peer k now report an excess vote of -2, causing $\Delta_{i,k}$ and Δ_i to become 0. Since by now Δ_i is less extreem than both $\Delta_{i,j}$ and $\Delta_{i,\ell}$ peer i is obliged to send to peers j and ℓ the excess votes it got from all other peers (a total of one excess vote to j and -2 to ℓ). After it had done so, in Figure 1(d), a new steady state is arrived.

6   Summary and Vision

Ubiquitous knowledge discovery poses challenges of two kinds. Some challenges will probably be answered by further development of existent technologies: The communication infrastructure, for instance, of both grid systems and peer-to-peer networks is expected to be further developed to facilitate tighter cooperation between different resources. Algorithms will certainly be developed which compute further primitives and at greater efficiency, etc.

The more interesting challenges, however, are those which require deeper inspection of the meaning and uses of knowledge in a ubiquitous system. We envision future work which will address problems of knowledge correctness and even coherence in ways which today are not familiar in the knowledge discovery community. We also anticipate the emergence of autonomous management technologies in ubiquitous systems, which we expect to rely to a great length on knowledge discovery. It is nevertheless important to remember that technology development in ubiquitous system is driven by the capacities of the devices and the needs of the applications. The further development of these will determine the course of development for ubiquitous knowledge discovery technologies.

References

1. Tanenbaum, A.S., Steen, M.V.: Distributed Systems: Principles and Paradigms, 2nd edn. Prentice-Hall, Englewood Cliffs (2007)
2. Litzkow, M.J., Livny, M., Mutka, M.W.: Condor - A hunter of idle workstations. In: Proc. of the International Conference on Distributed Computing Systems (ICDCS) (1988)
3. Johnston, W., Gannon, D., Nitzberg, B.: Grids as production computing environments: The engineering aspects of nasa's information power grid. In: Proc. 8th IEEE Symposium on High Performance Distributed Computing. IEEE Press, Los Alamitos (1999)
4. AlSairafi, S., Emmanouil, F.S., Ghanem, M., Giannadakis, N., Guo, Y., Kalaitzopoulos, D., Osmond, M.. Rowe, A., Syed, J., Wendel, P.: The design of discovery net: Towards open grid services for knowledge discovery. International Journal of High Performance Computing Applications 17(3), 297–315 (2003)
5. Brezany, P., Hofer, J., Tjoa, A., Wohrer, A.: Gridminer: An infrastructure for data mining on computational grids. In: APAC Conference and Exhibition on Advanced Computing, Grid Applications and eResearch (2003)
6. DataMiningGrid: EU Data Mining Grid Project (2007),
   http://www.datamininggrid.org/
7. Stankovski, V., Swain, M., Kravtsov, V.T., Niessen, D.W., Roehm, M., Trnkoczy, J., May, M., Franke, J., Schuster, A., Dubitzky, W.: Digging deep into the data mine with datamininggrid. IEEE Internet Computing (2008)
8. Palatin, N., Leizarowitz, A., Schuster, A., Wolff, R.: Mining for misconfigured machines in grid systems. In: Proceedings of the 12th ACM SIGKDD International Conference on Knowledge Discovery and Data Mining, pp. 687–692 (2006)
9. Chen, M., Zheng, A., Lloyd, J., Jordan, M., Brewer, E.: Failure diagnosis using decision trees. In: Proceedings of the International Conference on Autonomic Computing, New York, pp. 36–43 (2004)

10. Gu, L., Jia, D., Vicaire, P., Yan, T., Luo, L., Tirumala, A., Cao, Q., He, T., Stankovic, J.A., Abdelzaher, T., Krogh, B.H.: Lightweight detection and classification for wireless sensor networks in realistic environments. In: Proceedings of the 3rd International Conference on Embedded Networked Sensor Systems, pp. 205–217 (2005)

11. Madden, S.R., Franklin, M.J., Hellerstein, J.M., Hong, W.: Tinydb: an acquisitional query processing system for sensor networks. ACM Trans. Database Syst. 30(1), 122–173 (2005)

12. Rabbat, M.G., Nowak, R.D.: Decentralized source localization and tracking. In: Proceedings of the IEEE International Conference on Acoustics, Speech and Signal Processing, vol. 3, pp. 921–924 (2004)

13. Hoogen, I.V.D.: Deutsch's fallacies, 10 years after (2004), http://java.sys-con.com/read/38665.htm

14. Roberts, P.: Update: Hackers breach supercomputer centers. ComputerWorld news story (2004), http://www.computerworld.com/securitytopics/security/story/0,10801,92230,00.html

15. Datta, S., Kargupta, H.: Uniform data sampling from a peer-to-peer network. In: ICDCS 2007: Proceedings of the 27th International Conference on Distributed Computing Systems, Washington, DC, USA, p. 50. IEEE Computer Society, Los Alamitos (2007)

16. Wolff, R., Schuster, A.: Mining association rules in peer-to-peer systems. IEEE Transactions on Systems, Man and Cybernetics – Part B 34(6), 2426–2438 (2004)

17. Birk, Y., Liss, L., Schuster, A., Wolff, R.: A Local Algorithm for Ad Hoc Majority Voting via Charge Fusion. In: Guerraoui, R. (ed.) DISC 2004. LNCS, vol. 3274, pp. 275–289. Springer, Heidelberg (2004)

18. Krivitski, D., Schuster, A., Wolff, R.: A local facility location algorithm for sensor networks. In: International Conference on Distributed Computing in Sensor Systems (DCOSS), pp. 368–375 (2006)

19. Luo, P., Xiong, H., Lü, K., Shi, Z.: Distributed classification in peer-to-peer networks. In: KDD 2007: Proceedings of the 13th ACM SIGKDD International Conference on Knowledge Discovery and Data Mining, pp. 968–976. ACM, New York (2007)

20. Wolff, R., Bhaduri, K., Kargupta, H.: Local l2-thresholding based data mining in peer-to-peer systems. In: Proceedings of the Sixth SIAM International Conference on Data Mining (2006)

21. Branch, J., Szymanski, B., Gionnella, C., Wolff, R., Kargupta, H.: In-network outlier detection in wireless sensor networks. In: IEEE International Conference on Distributed Computing Systems, ICDCS 2006 (2006)

Resource Aware Distributed Knowledge Discovery

João Gama[1,2] and Antoine Cornuéjols[3]

[1] Faculty of Economics,
University of Porto, Portugal
[2] LIAAD - INESC Porto LA
University of Porto, Portugal
[3] AgroParisTech, Department MMIP
Paris, France
jgama@liaad.up.pt, antoine.cornuejols@lri.fr

1   The Challenge of Ubiquitous Computing

In the introduction it was argued that ubiquitous knowledge discovery systems have to be able to sense their environment and receive data from other devices, to adapt continuously to changing environmental conditions (including their own condition) and evolving user habits and need be capable of predictive self-diagnosis. In the last chapter, resource constraints arising from ubiquitous environments have been discussed in some detail. It has been argued that algorithms have to be *resource-aware* because of real-time constraints and of limited computing and battery power as well as communication resources.

In this chapter we discuss in some detail the implications that arise from this observation for the learning algorithms developed in ubiquitous knowledge discovery. We will argue that it implies a large shift from the current mainstream in data mining and machine learning.

While it is true that the last twenty years or so have witnessed large progress in machine learning and in its capability to handle real-world applications, machine learning so far has mostly centered on *one-shot data analysis from homogeneous and stationary data, and on centralized algorithms.* Nowadays we are faced with a tremendous amount of distributed data that could be generated from the ever increasing number of smart devices. In most cases, this data is transient, and may not be stored in permanent relations. In that context data mining approaches involving fixed training sets, static models and evaluation strategies are obsolete. A large part of the theory of machine learning (e.g. PAC learning model), which relies on the assumption that the data points are independent and identically distributed (meaning that the underlying generative process is stationary) has to be reconsidered and extended.

1.1   A World in Movement

The constraints we have enumerated imply to switch *from one-shot single-agent learning tasks to a lifelong and spatially pervasive perspective.* In the novel life-long perspective induced by ubiquitous environments, *finite training sets, static*

M. May and L. Saitta (Eds.): Ubiquitous Knowledge Discovery, LNAI 6202, pp. 40–60, 2010.
© Springer-Verlag Berlin Heidelberg 2010

models, and stationary distributions will have to be completely thought anew. These aspects entail new characteristics for the data:

- Data are made available through *unlimited streams* that continuously flow, eventually at high-speed, over time.
- The underlying *regularities may evolve over time* rather than be stationary.
- The data can no longer be considered as *independent and identically distributed*.
- The data is now often *spatially as well as time situated*.

Table 1. Forecast of electric load as example for a ubiquitous knowledge discovery application

An Ilustrative Problem: Forecast of Electrical Load. Electricity distribution companies usually set their management operators on SCADA/DMS products (Supervisory Control and Data Acquisition / Distribution Management Systems). One of their important tasks is to forecast the electrical load (electricity demand) for a given sub-network of consumers. Load forecast is a relevant auxiliary tool for operational management of an electricity distribution network, since it enables the identification of critical points in load evolution, allowing necessary corrections within available time, and planning strategies for different horizons[1].

In this application, data is collected from a set of sensors distributed all around the network. Sensors can send information at different time scales, speed, and granularity. Data continuously flow possibly at high-speed, in a dynamic and time-changing environment. Data mining in this context requires a continuous processing of the incoming data monitoring trends, and detecting changes. Traditional one-shot systems, memory based, trained from fixed training sets and generating static models are not prepared to process the high detailed data available, they are not able to continuously maintain a predictive model consistent with the actual state of the nature, nor are they ready to quickly react to changes. Moreover, with the evolution of hardware components, these sensors are acquiring computational power. The challenge will be to run the predictive model in the sensors itself.

But does the existence of ubiquitous environments really change the problem of machine learning? Wouldn't simple adaptations to existing learning algorithms suffice to cope with the new needs described in the foregoing? These new concerns might indeed appear rather abstract, and with no visible direct impact on machine learning techniques. Quite to the contrary, however, **even very basic operations that are at the core of learning methods are challenged in the new setting**. For instance, consider the standard approach to cluster variables (columns in a work-matrix). In a batch scenario, where all data is available

---

[1] A more detailed description appears in Chapter 7 Section 4.

and stored in a working matrix, we can apply any clustering algorithm over the *transpose* of the working matrix. In a scenario where data evolves over time, this is not possible, because the transpose operator is a blocking operator [1]: the first output tuple is available only after processing all the input tuples. Now, think of the computation of the entropy of a collection of data when this collection comes as a data stream which is no longer finite, where the domain of variables can be huge, and where the number of classes of objects is not known a priori; or think on continuous maintenance the k-most frequent items. And then, what becomes of statistical computations when the learner can only afford a one pass on each data piece because of time and memory constraints; when he has to decide on the fly what is relevant and must be further processed and what is redundant or not representative and could be discarded? These are but a few examples of a clear need for new algorithmic approaches in the KDubiq framework. Table 1 illustrates the problem from an application perspective.

Data mining is thus faced with a set of new challenges that all share common issues. Namely, the data comes from distributed sources in continuous flows generated by evolving distributions; the description space (set of attribute-values) can be huge and even unbounded; computation resources (processing power, storage capacity, communication bandwidth, battery power) are limited.

In short, machine learning algorithms will have to enter the world of **limited rationality**, e.g. rational decisions are not feasible in practice due to the finite computational resources. To reshape, ubiquitous data mining implies new requirements to be considered. The new constraints include:

- The algorithms will have to use *limited computational resources* (in terms of computations, space and time).
- The algorithms will have only a *limited direct access to data* and may have to communicate with other agents on *limited bandwidth* resources.
- In a community of smart devices geared to ease the life of users in real time, answers will have to be ready in an *anytime protocol*.
- Overall, data gathering and data (pre-)processing will be *distributed*.

In this chapter, we discuss algorithmic issues related to advanced data analysis in dynamic environments using devices with limited resources. In the contexts we are interested in, data is distributed, flowing possibly at high-speed. We identify the limitations of the current machine learning theory and practice and underline the most desirable characteristics for a learning algorithm in these scenarios. The analysis follows three main dimensions: learning in resource aware devices, distributed data, and continuous flow of data. These dimensions define the organization of the chapter.

2   The Challenge of Limited Resources

Ubiquitous environments mark the end of the area of the single and unlimited learning agent paradigm. The main point here is the need for a high *degree of autonomy* of the agents. We would like to maximize the level of autonomy, requiring the ability to process their own data, and make local decisions. Most important, agents might communicate with neighbors, minimizing communication costs, to make collective and rational decisions. The fact that a multitude of simple and situated agents are now to form a new kind of anytime responsive system implies a profound reexamination of the data processing and data mining tasks.

Traditional data mining applications call for an off-line centralized data analysis processing over an existing (very) large database. In ubiquitous data mining (UDM), every one of the assumptions that underlie this picture are challenged. Indeed, in this setting, *data is transient*, obtained through a web of, possibly heterogeneous, sensors with limited computational, memory and communication bandwidth capabilities. Furthermore, both for reasons of scarce resources and because of real-time demands, data must be processed, at least partly, on the fly and locally. Applications in fusion of environmental sensor measures, in the analysis of information gathered by web/blog crawlers that independently sip through collections of texts and links, in analysis and control of distributed communication networks have been already cited. But many more are coming fast in scientific, business and industrial contexts.

These new constraints challenge the existing data mining techniques and require a significant research effort to overcome their current limitations.

First of all, *at the level of each individual device or agent*, the limitations on both computing power, memory space and time will constrain the type of processing that can be done. For instance, floating point operations may be unavailable, and processing power may be severely limited beyond the one required to run the normal operations of the device. Therefore, algorithms for local data mining must require very low complexity processes, possibly at the cost of degraded output. In many settings, sensor nodes and computing units (e.g. handheld devices) lack sufficient memory to run classical data mining techniques which require that the results of the intermediate processing steps be resident in memory over the processing time of the running algorithm. Data, often coming in the form of continuous streams, will have either to be partitioned into subsets small enough as to be tractable, or processed on a one-pass only basis. Furthermore, the locally available data may be too restricted to permit an informed enough data mining process, and queries will have to be addressed to other agents. Similarly, it is possible to envision cases where one device will have to recruit the computing power of other less busy agents to meet its own needs. In all these cases, this will require that the individual devices have sufficient self-monitoring capabilities so as to know when to ask for additional information or for help in the form of computing power.

Secondly, *at a more integrated level*, several issues will need to be addressed. To begin with, at quite a basic level, the heterogeneity of the individual devices will introduce problems of communication when the set of measurements and the formats in which they are expressed will differ among agents. For instance, in sensor networks, work on a sensor model language called sensorML has been started as part of the Opengeospatial Consortium. It uses XML to encode and describe sensor models. The whole process ranging from input to output and associated parameters is described using this language. Apart from the issue of heterogeneousness of the devices, the whole system will have to be responsive even in face of faulty or missing measurements, drifting characteristics of the sensors and generally haphazardous malfunctions. Furthermore, measured data might well be redundant on one side, but also different in their own right because issued from different locations and aimed at different tasks. These characteristics will require that methods for the management of the system be developed. With regards to learning, meta-learning methods will be needed in order to distribute learning processes when needed and to integrate the results on a anytime basis. Since the architecture of the system may well not incorporate a global master device, these management capabilities will have themselves to be distributed and implemented as parts of the self-aware computing capacities of the individual agents. In this respect, much remains to be done, and we do not have, at this time, a mature methodology to meet that challenge.

The limited communication bandwidth between the agents poses other questions. In-network knowledge integration is an open research issue in ubiquitous data mining. It has impact on many problems like: merging of clustering models, of classification decisions and frequent patterns mining. Controlling the accuracy of the resulting integrated knowledge is fast becoming a critical research question. Exchange of data should not be done without some form of preprocessing that clean the data, selects a right degree of precision, remove outliers and generally filter out irrelevant or non representative measurements. Techniques exist for preprocessing data in data mining, however, they often assume that data is available in its entirety and that sufficient computing power and memory space is available. As will be seen in more details in the sections devoted to data streaming, algorithms have to be completely reconsidered in the limited rationality and lifelong setting. At the other end of the learning process, the results of learning that have to be stored or shared between agents should require only compact transmissions and storage.

Overall, new techniques for intelligent sampling, to use generate synopsis and summarized information, and to output approximate solutions are required. But this brings a new problem. Since the degree of approximation is dependent on the specifics of the ubiquitous system and on the application, ubiquitous data mining systems must be able to adapt to the characteristics of the problem and to their own limitations. In other words, they must be *self-aware* to a certain degree in order to optimize their own parameters and deliver on-time the best possible analysis. In the field of data mining per se, a powerful meta-learning approach has been developed since the 90s, that of *ensemble learning* [2]. The technique makes use of the combination of multiple models learned by *weak learners* in order to *boost* the overall performance of the learning system. In the context of ubiquitous data mining, it is tempting to view these ensemble learning techniques as means to turn the curse of distributed and limited data mining algorithms into a strength, by using local models as base learners. Nevertheless, so far, few works [3] have addressed the problem of incremental, online, and distributed maintenance of ensembles of decision models [4].

3   The Challenge of Distributed Systems

Mining spatially tagged data. There are now several data generation environments where the nodes generating the data are spatially spread and interrelated and where these relations are meaningful and important for data mining tasks. New types of applications are thus emerging in health services, in environmental monitoring or in distributed and real-time control of vehicles. It is also possible to record characteristics of the landscape itself as in satellite recordings and to monitor the spatial location of objects. The latter has become very popular with the emergence of mobile phones, GPS and RFID. In sensor networks, for instance, sensors are often explicitly spatially related to each other and this spatial information can be considered as a king of meta tagging of the data. In other applications, where, for instance, the agents themselves are moving (e.g. vehicles on roads), GPS information can accompany the set of measurements made.

In all of these applications, the spatial information is essential in the processing of the data and the discovery of meaningful regularities. So far, works on spatial data mining has been scarce even though applications in satellite remote sensing and geographical databases have spurred a growing interest in methods and techniques to augment databases management and mining with spatial reasoning capabilities.

The combination of distributed systems and the pervasive/ubiquitous computing paradigms can lead into the area of distributed data mining. It is a broad area characterized by combinations of multiple learning algorithms as well as multiple learning problems. These learning problems are also directed by constraints such as security, privacy, communication costs and energy restrictions. The idea of ubiquitous knowledge discovery and autonomous agents is closely related. An agent needs to be able to discover, sense, learn, and act on its own. There are basically two types of agent architectures, one in which the agents are tied to one node and one in which the agents move between nodes. In systems such as Objectspace's *Voyager*, the agents move between the machines, while a peer-to-peer system such as *JXTA* only allows agents to be tied to the nodes. In both systems the agents either reside or migrate to nodes that are prepared for receiving. Distributed objects can be seen as a premise for agents: only some additional characteristics need to be adorned to the distributed object [5]. The key issues in conceiving and implanting intelligent agents is the agent's learning model, coordination and collaboration.

One example of interaction between the agents is the *Java Agents for Meta-Learning* (JAM) system developed by Stolfo et al. [6]. The agent can communicate with other agents using the *Knowledge Query and Manipulation Language*

(KQML) [7] together with the Knowledge Interchange Formalism (KVI) [8] agent information exchange format. These communications and associated exchange formats allow the agent to express first order requests. A complementary system for agent corporation is the *Open Agent Architecture* (OOA) [9]. In some cases the agents will query other agents and perhaps exchange examples or models. To exchange an example sounds simple, but it requires that the data format is agreed upon and understood on both sides of the exchange. Research in general purpose languages for model exange is recent. An example is the Predictive Model Markup Language (PMML), a XML mark up language to describe statistical and data mining models.

The strong limitations of the centralized approach is discussed in [10]. The authors point out *'a mismatch between the architecture of most off-the-shelf data mining algorithms and the needs of mining systems for distributed applications'.* Such mismatch may cause a bottleneck in many emerging applications. Most important, in applications like monitoring, centralized solutions introduce delays in event detection and reaction, that can make mining systems useless.

3.1   Sensor Networks

The advent of widely available and cheap computer power, in parallel with the explosion of networks of all kinds (wired and wireless), opens new unknown possibilities for the development and self-organization of communities of intelligent communicating appliances. Sensor networks are one such class of distributed environments that have the potential to redefine the way data gathering and data intensive applications will be carried out in the future.

Sensor networks can be used to detect the occurrence of complex events patterns with spatial and temporal characteristics[2]. Most of these sensor networks deal with distributed and constrained computing, storage, power, and communication-bandwidth resources[3]. Traditionally, sensor networks communicate sensed data to a central server that runs offline data mining algorithms. In the KDubiq perspective, analysis should be done *in situ* using local information. Mining in such environments naturally calls for proper utilization of these distributed resources. Restrictions on the size and power source will define how complex is the processing performed in the sensors. We may have, at one end of the spectrum, centralized systems and, at the other end, completely distributed systems, where almost all the intelligent processing is performed in the sensors. The solution considered in Ubuiquitous Data Mining can broadly be seen as belonging to one of the following categories: process-based, data-based, or network-based.

The focus of *process-based* approaches is to deal with the actual processing of the algorithm. Approximation and randomized algorithms are solutions used in this category [11], to generate approximate but compact summaries and data synopsis [12]. Approximation and randomization are used together with sliding

---

[2] See Chapter 4, Section 3 for more details.
[3] See Chapter 2 for more details.

windows in a two-fold objective. The first is to capture the most recent output. The other objective is to solve the problem of concept drift. *Data-based* solutions are used whenever data flows at high-speed, faster than processing capabilities. Data can be *slow downed* before the actual learning phase takes place. It could be in the form of processing a subset of data using techniques such as *sampling* and *load shedding*. It also could be in the form of dimensionality reduction such as *sketching*. In *network-based solutions*, the data mining algorithm is executed locally onboard of a sensor node. The algorithm allows message exchanges in order to integrate local results. Approximate global models could be computed accordingly. Examples of networks include peer-to-peer (P2P) computing [13], clustering of sensor nodes and efficient routing [14], and grid computing [15]. They are discussed in detail in Chapter 2. Some examples of applications include clustering of sensor nodes aims at prolonging the network life-time by choosing nodes with highest energy to be the cluster heads. Cluster heads then collect and aggregate data to be sent to the base station. In grid computing, a so-called knowledge grid is built to benefit from the services provided by the grid in the distributed data mining process. These services include information and resource management, communication and authentication.

Another problem that emerge, in many ubiquitous applications, where information must be exchanged between nodes are privacy issues. This aspect is discussed in detail in Chapter 5.

Data mining in sensor networks includes all the ingredients of KDubiq perspective: distributed computation, with time, space and resources bounds. Current data mining techniques assume iid (independently and identically distributed) examples, static models, finite training sets, and unrestricted resources. Furthermore, they usually ignore resources constraints, entailing high maintenance costs: for instance retraining models from time to time. By contrast, the KDubiq approach, associated with analysis *in situ*, maintenance of dynamic models that evolve over time, change detection mechanisms, awareness of the available resources, is the most appropriate option.

Data stream mining techniques use the above approaches in different data mining paradigms and tasks including clustering, classification, frequent pattern discovery, time series analysis and change detection [17]. While few of these techniques are resource-aware when applying the above solution approaches, others are not. This problem needs to be addressed in order to realize UDM algorithms in real-life applications. For example, a data stream mining technique often uses a linear memory space to store past observations. Without precaution, this may overwhelm the limited available memory in a ubiquitous environment. However, if the algorithm is resource-aware, the approximation factor can change over time in order to cope with the critically low availability of memory. The same analogy applies to the different available resources. This has been demonstrated in [18]. The area of adaptation and resource-awareness is still open. Many data stream mining algorithms are required to be resource-aware and adaptive in order to realize its applicability in ubiquitous computing environments.

Table 2. Description of a distributed clustering algorithm

An Illustrative Algorithm: Distributed Clustering. The scenario is in a sensor network where each sensor produces a continuous stream of data. Suppose we have m distributed sites, and each site i has a data source S_i^t at time t. The goal is to continuously maintain a k-means clustering of the points in $S^t = \cup_i^m S_i^t$. The distributed algorithm presented in *Conquering the Divide: Continuous Clustering of Distributed Data Streams* [16] is based on the *Furthest Point* clustering.

Furthest Point Clustering. The base idea consists in selecting randomly the first cluster center c_1 among data points. Subsequent $k-1$ cluster centers are chosen as the points that are more distant from the previous centers c_1, c_2, ..., c_{i-1}, by maximizing the minimum distance to the centers. This algorithm requires k passes over training points. It has an interesting property. It ensures a 2-approximation of the optimal clustering. A skeleton of the proof is: Suppose that the $k+1$ iteration produces $k+1$ points separated by a distance at least D. The optimal k clustering must have a diameter at least D. By the triangular inequality the chosen clustering has diameter at most 2D.

The Parallel Guessing Clustering. Based on the *Furthest Point* algorithm the authors developed a one pass clustering algorithm: the *Parallel Guessing Clustering.* The idea basically consists in selecting an arbitrary point as the first center, and for each incoming point p compute $r_p = min_{c \in C} d(p, c)$. If $r_p > R$, set $C = C \cup p$. This strategy would be correct if we knew R, however, in unbounded data streams, R is unknown in advance! The solution proposed in [16] consists in making multiple guesses for R as $(1+\epsilon/2)$, $(1+\epsilon/2)^2$, $(1+\epsilon/2)^3$, and run the algorithm in parallel. If a guess for the value of R generates more than k centers, this imply that R is smaller than the optimal radius (R_o). When $R \geq 2R_0$ and k or less centers are found then a valid clustering is generated.

Conquering the Divide Each local site maintains a *Parallel Guessing Algorithm* using its own data source. Whenever it reaches a solution, it sends to the coordinator the k centers and the radius R_i. Each local site only re-send information when the centers change. The coordinator site maintains a *Furthest Point Algorithm* over the centers sent by local sites.

4   The Challenge of Streaming Data

The time situation of data. The fact that, in ubiquitous environments, data are produced on a real-time basis, or, at least, in a sequential fashion, and that the environment and the task at hand may change over time, profoundly modifies the underlying assumptions on which rest most of the existing learning techniques and demands the development of new principles and new algorithms.

In the following, we discuss these new set of issues along two main directions. The first one deals with the sequential nature of data, the fact that it comes as streams of indefinite length and must be processed with limited resources. The second ones discusses the consequences of the fact that data can no longer be considered as independently and identically distributed *iid* for short.

In the last two decades, machine learning research and practice has focused on batch learning usually with small datasets. In batch learning, the whole training data is available to the algorithm, that outputs a decision model after processing the data eventually (or most of the times) multiple times. The rationale behind this practice is that examples are generated at random accordingly to some stationary probability distribution. Most learners use a greedy, hill-climbing search in the space of models.

4.1   Static versus Streaming

What distinguishes current data sets from earlier ones are the continuous flow of data and the automatic high-speed data feeds. We do not just have people who are entering information into a computer. Instead, we have computers entering data into each other [11]. Nowadays there are applications in which the data is best modeled not as persistent tables but rather as transient data streams. In some applications it is not feasible to load the arriving data into a traditional DataBase Management Systems (DBMS), and traditional DBMS are not designed to directly support the continuous queries required in these applications [12].

Algorithms that process data streams must provide fast answers using little memory resources. But this is usually at the cost of delivering approximate solutions. The requirement of an exact answer is relaxed to a criterion where the guarantee is on delivering an approximate answer within a small error range and with high probability. In general, as the range of the error decreases the needs in computational resources goes up. In some applications, mostly database oriented, it is required that the approximate answer be within an admissible error margin. In this case, results on tail inequalities provided by statistics are useful to accomplish this goal. The basic general bounds on the tail probability of a random variable (that is, the probability that a random variable deviates greatly from its expectation) include the Markov, Chebyshev and Chernoff inequalities [19].

A set of techniques have been developed in Data Streams Management Systems that allow one to store compact stream summaries that are nonetheless guaranteed to approximately solve queries. All these approaches require controlling a trade-off between accuracy and the amount of memory used to store the summaries, under the additional constraint that a limited time only is allocated to process data items [11]. It is observed that the most common problems end up computing quantiles, frequent item sets, and storing frequent counts along with error bounds on their true frequency.

From the viewpoint of a data streams management system, several research issues emerge. For instance, one such issue is related to the need of approximate query processing techniques in order to evaluate queries that require unbounded amount of memory. Sampling techniques have been used to handle situations where the flow rate of the input stream is faster than the query processor. One important question is raised by the existence of blocking operators (e.g. aggregation and sorting) in the presence of unending streams. It is essential to identify them and to find ways to circumvent their blocking effect. The following table summarizes the main differences between traditional and stream data processing:

	Traditional	Stream
Number of passes	Multiple	Single
Processing Time	Unlimited	Restricted
Available Memory	Unlimited	Fixed
Result	Accurate	Approximate
Distributed	No	Yes

4.2   When Data Points Are No Longer IID

When the samples of data are both spatially and time situated, data points can no longer be considered as independently and identically distributed, a fact that is compounded when the underlying generative process is itself changing over time.

One consequence is that the statistical theory of learning [20] does not hold anymore. In particular, the inductive criteria that are based on additive measures of cost (e.g. the empirical risk and the real risk associated with a candidate hypothesis) are no longer satisfactory. The question then becomes: how to replace this fundamental theory? Works on weak long term correlations exist in statistics, but they are still far from answering the challenge that faces machine learning in this respect.

One interesting question concerns ordering effects: the fact that the result of a (on-line) learning session may depend on the order of presentation of the data points. This has usually been considered as a nuisance that one should try to get rid of. However, the order of the data points can clearly convey potentially useful information about the *evolution* of the underlying generative process, or even, this order could result from a clever and helpful teacher. In this case, one should on the contrary try to take advantage from this source of information. Apart from a few works in constructive induction and some pioneering works in inductive logic programming, almost everything remains to be done in this direction. Reasoning about the evolution of the learning process is also a potential research line that remains almost unexplored.

Concretely, data streams require that the learners be endowed with on-line or incremental learning capabilities. This means in particular that they should be able to process the data sequentially as they arrive and that they should be able to produce either decisions or hypotheses about the surrounding world in an anytime fashion. Since, their memory is limited and because the underlying generative process, or the task at hand, may change, they have to be able to measure the relevance of a piece of information over time, and be able to *forget* no longer relevant data pieces.

Questions about how to sample and summarize the data, how to carry on on-line learning are discussed below.

4.3   Where We Are

Recent developments in machine learning point out directions for learning in ubiquitous environments.

Approximation and Randomization. Approximation and Randomization techniques has been used to solve distributed streaming learning problems. Approximation allows one to get answers that are correct within some fraction ϵ of error, while *randomization* allows a probability δ of failure. The base idea consists of mapping a very large input space to a small synopsis of size $O(\frac{1}{\epsilon^2}log(\frac{1}{\delta}))$. Approximation and Randomization techniques has been used to solve problems like measuring the entropy of a stream, association rule mining frequent items [21], k-means clustering for distributed data streams using only local information [16], etc.

Learning from Infinite Data in Finite Time. For example, in [24] the authors present a general method to learn from arbitrarily large databases. The method consists in, first, deriving an upper bound for the learner's loss as a function of the number of examples used in each step of the algorithm. It then uses this to minimize each step's number of examples, while guaranteeing that the model produced does not differ significantly from the one that would be obtained with infinite data. This general methodology has been successfully applied in k-means clustering [24], decision trees [23,25,26], etc. We have pointed out the necessity of new algorithms for clustering variables in the stream setting. The Online Divisive-Agglomerative Clustering (ODAC) system [27] continuously maintains a tree-like hierarchy of clusters that evolves with data. ODAC continuously monitor the evolution of cluster's diameters defined as the distance between the farthest pair of streams. It uses a *splitting operator* that divides an existing cluster into two new more detailed clusters as response to more data is available, and a *merge operator* that agglomerates two sibling clusters, in order to react to changes in the correlation structure between time series. In stationary environments expanding the structure leads to a decrease in the diameters of the clusters. The system is designed to process thousands of data streams that flow at high-rate. The main features of the system include update time and memory consumption that do not depend on the number of examples in

Table 3. An algorithm for counting from data streams

An Illustrative Algorithm: Counting from Data Streams. The Count-Min Sketch [22] is a streaming algorithm for summarizing data streams. It has been used to approximately solve point queries, range queries, and inner product queries. Here we present a simple point query estimate: how many times an item in high-speed stream as been seen so far?

Data Structure. A Count-Min Sketch is an array of $w \times d$ in size. Given a desired probabilty level (δ), and an admissible error (ϵ), $w = 2/\epsilon$ and $d = log(1/\delta)$.

Update Procedure. Each entry x in the stream, is mapped to one cell per row of the array of counts. It uses d hash functions to map entries to $[1..w]$. When an update c of item j arrives, c is added to each of these cells.

Computing Point Queries. At any time, the estimate $\hat{x}[j]$ is given by taking $min_k CM[k, h_k(j)]$.

Properties. The important properties of this estimate are: $x[j] \leq \hat{x}[j]$ and $\hat{x}_i \leq \epsilon \times ||x_i||_1$, with probability $1 - \delta$.

Table 4. Decision tree learning from data streams

The basic algorithm for learning a decision tree from a data stream. The Very Fast Decision Tree algorithm presented in [23]:

- **Input:** δ desired probability level; τ Constant to Solve Ties.
- **Output:** T A decision Tree
- **Init:** $T \leftarrow$ Empty Leaf (Root)
- While (TRUE)
  - Read next Example
  - Propagate Example through the Tree from the Root till a leaf
  - Update Sufficient Statistics at leaf
  - Time to Time
    /* after processing a minimum number of examples */
    * Evaluate the merit $G(.)$ of each attribute
    * Let A_1 the best attribute and A_2 the second best
    * Let $\epsilon = \sqrt{R^2 ln(1/\delta)/(2n)}$
    * If $G(A_1) - G(A_2) > \epsilon$ OR $(G(A_1) - G(A_2) < \epsilon < \tau)$
    *   Install a splitting test based on A_1
    *   Expand the tree with two descendant leaves

the stream. Moreover, the time and memory required to process an example decreases whenever the cluster structure expands.

The proposed method can be useful to solve one aspect of Learning in ubiquitous environments. Nevertheless many others remain unsolved. For example, in order to learn from distributed data, we need efficient methods in minimizing the communication overheads between nodes. Work in this direction appears in [28].

The idea of the agent's *limited rationality* leads us to ensemble models used to collectively solve problems. In [4], the authors proposed a method that offer an effective way to construct a redundancy-free, accurate, and meaningful representation of large decision-tree ensembles often created by popular techniques such as Bagging, Boosting, Random Forests and many distributed and data stream mining algorithms.

4.4   Online, Anytime and Real-Time Learning

The challenge problem for data mining is the ability to continuously maintain an accurate decision model. In this context, the assumption that examples are generated at random according to a stationary probability distribution does not hold, at least in complex systems and for large periods of time.

In the presence of a non-stationary distribution, the learning system must incorporate some form of forgetting past and outdated information. Learning from data streams require incremental learning algorithms that take into account these changes in the data generating process. Solutions to these problems require new sampling and randomization techniques, and new approximate, incremental and decremental algorithms. In [24], the authors identify desirable properties of learning systems that are able to mine continuous, high-volume, open-ended data streams as they arrive. Learning systems should be able to process examples and answering queries at the rate they arrive. Overall, some desirable properties for learning in data streams include: incrementality, online learning, constant time to process each example, single scan over the training set, and taking drift into account.

Cost-Performance Management. Incremental learning is one fundamental aspect for the process of continuously adaptation of the decision model. The ability to update the decision model whenever new information is available is an important property, but it is not enough. Another required operator is the ability to *forget* past information [29]. Some data stream models allow delete and update operators. Sliding windows models require forgetting old information. In all these situations the incremental property is not enough. Learning algorithms need forgetting operators that reverse learning: decremental unlearning [30].

The incremental and decremental issues requires a continuous maintenance and updating of the decision model as new data is available. Of course, there is a trade-off between the cost of update and the gain in performance we may obtain [31]. The update of model depends of the complexity of its representation language. For instance, very simple models, using few free-parameters, are easy to adapt in face of changes. Small and potentially insignificant variations in the data will not result in unwarranted changes in the learned regularities. The variance of the hypothesis class is said to be limited. However, the downside is an associated high bias. The hypothesis space cannot accommodate a large variety of data dependencies. This is known as the bias-variance tradeoff. This tradeoff is fundamental in classical one-shot learning. It becomes even more so in on-line learning. In addition, the more complex the hypothesis space (the larger the number of effective parameters), the more costly are the update operations. Effective means of controlling it must be devised. Continuous learning therefore requires efficient control strategies in order to optimize the trade-off between the gain in performance and the cost of updating. To this aim, incremental and decremental operators have been proposed, including sliding windows methods.

An illustrative example is the case of ensemble learning techniques, such as boosting, that rely on learning multiple models. Theoretical results show that it is thus possible to obtain arbitrary low errors by increasing the number of models. To achieve a linear reduction of the error, we need an exponential increase in the number of models. Finding the *Breakeven* point, that is the point where costs equalize benefits, and not going behind that, is a main challenge.

Monitoring Learning. When data flows over time, and, at least, in case of large periods of time, the assumption that the examples are generated at random according to a stationary probability distribution becomes highly unlikely. In complex systems and for large time periods, we should expect changes in the distribution of the examples. A natural approach for these *incremental tasks* are *adaptive learning algorithms*, incremental learning algorithms that take into account concept drift.

Concept drift means that the concept related to the data being collected may shift from time to time, each time after some minimum permanence. Changes occur over time. The evidence for changes in a concept are reflected in some way in the training examples. Old observations, that reflect the past behavior of the nature, become irrelevant to the current state of the phenomena under observation and the learning agent must forget that information.

The nature of change is diverse. Changes may occur in the context of learning, due to changes in hidden variables, or in the characteristic properties of the observed variables. In fact, it is usual to distinguish between:

- Changes in the underlying distribution over the instance descriptions (denoted $D_{\mathcal{X}}$, the distribution over the description space $\mathcal{X}$).
- Changes in the conditional distribution of the label w.r.t. the description (denoted $D_{\mathcal{Y}|\mathcal{X}}$, where $\mathcal{Y}$ is the label space). This is usually called "concept drift".
- Changes in both of the distributions.

In the former case, the change affects the space of the available instances, but not the underlying regularity. This, however, may require changes in the decision rules in order to optimize the decision process over the dense regions of the instance space. In the second case, the identified decision rule has to be adapted.

Most learning algorithms use blind methods that adapt the decision model at regular intervals without considering whether changes have really occurred. Much more interesting is explicit change detection mechanisms [32]. The advantage is that they can provide meaningful description (indicating change-points or small time-windows where the change occurs) and quantification of the changes. The main research issue is how to incorporate change detection mechanisms in the learning algorithm. Embedding change detection methods in the learning algorithm is a requirement in the context of continuous flow of data. The level of *granularity* of decision models is a relevant property, because if can allow partial, fast and efficient updates in the decision model instead of rebuilding a complete new model whenever a change is detected. The ability to recognize seasonal and re-occurring patterns is an open issue.

Novelty Detection. Novelty Detection refers to the automatic identification of unforeseen phenomena embedded in a large amount of normal data. It corresponds to the appearance of a new concept (e.g. a new label or a new cluster) *from unlabelled data.* Learning algorithms must then be able to identify and learn new concepts [33]. *Novelty* is always a relative concept with regard to our current knowledge. Intelligent agents that act in dynamic environments must be able to learn conceptual representations of such environments. Those conceptual descriptions of the world are always incomplete. They correspond to what is *known* about the world. This is the *open* world assumption as opposed to the traditional *closed* world assumption, where what is to be learnt is defined in advance. In open worlds, learning systems should be able to extend their representation by learning new concepts from the observations that do not match the current representation of the world. This is a difficult task. It requires identifying the *unknown*, that is, the limits of the current model. In that sense, the *unknown* corresponds to an *emerging pattern* that is different from *noise*, or *drift* in previously known concepts.

4.5   Issues and Challenges in Learning from Distributed Data Streams

Streaming data and domains offer a nice opportunity for a symbiosis between streaming data management systems and machine learning. The techniques developed to estimate synopsis and sketches require counts over very high dimensions both in the number of examples and in the domain of the variables. The techniques developed in data streams management systems can provide tools for designing machine learning algorithms in these domains. On the other hand, machine learning provides compact descriptions of the data than can be useful for answering queries in DSMS.

Incremental Learning and Forgetting. In most applications, we are interested in maintaining a decision model consistent with the current status of the nature.

This lead us to the sliding window models where data is continuously inserted and deleted from a window. Learning algorithms must have operators for incremental learning and forgetting. Incremental learning and forgetting are well defined in the context of predictive learning. The meaning or the semantics in other learning paradigms (like clustering) are not so well understood, very few works address this issue.

Change Detection. *Concept drift* in the predictive classification setting is a well studied topic. In other learning scenarios, like clustering, very few works address the problem. The main research issue is how to incorporate change detection mechanisms in the learning algorithm for different paradigms.

Feature Selection and Pre-processing. Selection of relevant and informative features, discretization, noise and rare events detection are common tasks in machine learning and data mining. They are used in a one-shot process. In the streaming context the semantics of these tasks changes drastically. Consider the feature selection problem. In streaming data the concept of *irrelevant* or *redundant* features are now restricted to a certain period of time. Features previously considered *irrelevant* may become *relevant*, and vice-versa to reflect the dynamics of the process generating data. While in standard data mining, an irrelevant feature could be ignored forever, in the streaming setting we need still to monitor the evolution of those features. Recent work based on the *fractal dimension* [34] could point towards interesting directions for research.

Ubiquity in the Feature Space. In the static case, similar data can be described with different schemata. In the case of dynamic streams, the schema of the stream can also change. We need algorithms that can deal with evolving feature spaces over streams. There is very little work in this area, mainly pertaining to document streams. For example, in sensor networks, the number of sensors is variable (usually increasing) over time. For instance, clustering of data coming through streams on different sensing and computing sites is a growing field of research that brings completely new algorithms (see for instance [35,36]).

Evaluation Methods and Metrics. An important aspect of any learning algorithm is the hypothesis evaluation criteria. Most of evaluation methods and metrics were designed for the static case and provide a single measurement about the quality of the hypothesis. In the streaming context, we are much more interested in how the evaluation metric evolves over time. Results from the *sequential statistics* [8] may be much more appropriate.

There is a fundamental difference between learning from small datasets and large datasets. As pointed-out by some researchers [38], current learning algorithms emphasize variance reduction. However, learning from large datasets may be more effective when using algorithms that place greater emphasis on bias management.

5   Where We Want to Go: Emerging Challenges and Future Issues

KDubiq refers to systems and algorithms to come which will exhibit high level of autonomy but will be subjected to resources constraints. These systems aim to address the problems of data processing, modeling, prediction, clustering, and control in changing and evolving environments. They should be able to self-evolve their structure and their knowledge on the environment.

The current data mining perspective assumes a world that is static and that we can monitor using unrestricted resources. In this approach, for instance, the arrival of new information can be accommodated simply by retraining on the updated data base. Limits on space and time complexity are not considered an issue. This is completely at odds with the KDubiq perspective in which **the world is dynamic and things evolve in time and space**, and where the **cognitive agents only enjoy limited rationality**. In this new challenging framework, learning requires decision models that evolve over time, mechanisms for change detection, and the ability to take into account the resources available.

Accordingly, all steps in the KDD process must be thought anew. A lot of the preprocessing stages and of the choices that usually were made prior to learning *per se* must now be considered as integral parts of the learning process that, furthermore, must be continuously monitored. In addition, any learning algorithm dedicated to knowledge discovery in ubiquitous systems must take memory, space, time, communication, and other costs into account. This represents a revolution in the way data mining algorithms are conceived. In this context, the relevant point *is not what we can do more, but what we can do better.*

The definition of *standards* to represent and exchange models and patterns, the design of techniques for the *incorporation of domain knowledge* and *visualization*, the *definition of processes and systemic approaches*, are traditional issues in data mining. Their importance is reinforced in distributed and network environments. Emerging applications like *bio-informatics, semantic Web, sensor networks, radio frequency identification*, etc. are all dependent upon the solution to these issues.

Simple objects that surround us are changing from static, inanimate objects into adaptive, reactive systems with the potential to become more and more useful and efficient. Smart things associated with all sort of networks offers new unknown possibilities for the development and self-organization of communities of intelligent communicating appliances. Learning in these contexts must be *pervasive*, and become invisible.

We are witnessing the emergence of systems and networked organizations with high-level of complexity. Nowadays many tasks, e.g. missions to Jupiter, are planed 10 or more years in advance, projecting to the future technologies not yet available. These systems and organizations must be capable of autonomously achieving desired behaviors. machine learning can make the difference, but learning systems must be able to autonomously define what and when to learn, define plans and strategies for learning, define the relevant goals, search and select the required information, minimize cost and risk, incorporate uncertainty, etc. KDubiq is a step towards this direction.

References

1. Barbar, D.: Requirements for clustering data streams. SIGKDD Explorations 3(2), 23–27 (2002)
2. Schapire, R.: Strength of weak learnability. Journal of Machine Learning 5, 197–227 (1990)
3. Oza, N.: Online Ensemble Learning. PhD thesis, University of California, Berkeley (2001)
4. Kargupta, H., Dutta, H.: Orthogonal Decision Trees. In: Proceedings of The Fourth IEEE International Conference on Data Mining (ICDM 2004), Brighton, UK (2004)
5. Davies, W., Edwards, P.: Agent-Based Knowledge Discovery. In: AAAI Spring Symposium on Information Gathering (1995)
6. Stolfo, S.J., Prodromidis, A.L., Tselepis, S., Lee, W., Fan, D.W., Chan, P.K.: JAM: Java agents for meta-learning over distributed databases. In: Knowledge Discovery and Data Mining, pp. 74–81 (1997)
7. Finin, T., Fritzson, R., McKay, D., McEntire, R.: KQML as an Agent Communication Language. In: Adam, N., Bhargava, B., Yesha, Y. (eds.) Proceedings of the 3rd International Conference on Information and Knowledge Management (CIKM 1994), Gaithersburg, MD, USA, pp. 456–463. ACM Press, New York (1994)
8. Genesereth, M.R., Fikes, R.E.: Knowledge Interchange Format, Version 3.0 Reference Manual. Technical Report Logic-92-1, Stanford University, Stanford, CA, USA (1992)
9. Martin, D., Cheyer, A., Moran, D.: The Open Agent Architecture: a framework for building distributed software systems. Applied Artificial Intelligence 13(1/2), 91–128 (1999)
10. Park, B., Kargupta, H.: Distributed Data Mining: Algorithms, Systems and Applications. In: Data Mining Handbook. Lawrence Erlbaum Associates, Mahwah (2002)
11. Muthukrishnan, S.: Data streams: algorithms and applications. Now Publishers (2005)
12. Babcock, B., Babu, S., Datar, M., Motwani, R., Widom, J.: Models and issues in data stream systems. In: Kolaitis, P.G. (ed.) Proceedings of the 21nd Symposium on Principles of Database Systems, pp. 1–16. ACM Press, New York (2002)
13. Datta, S., Bhaduri, K., Giannella, C., Wolff, R., Kargupta, H.: Distributed data mining in peer-to-peer networks. IEEE Internet Computing special issue on Distributed Data Mining 10(4), 18–26 (2006)
14. Younis, O., Fahmy, S.: Heed: a hybrid, energy-efficient, distributed clustering approach for ad hoc sensor networks. IEEE Transactions on Mobile Computing 3(4), 366–379 (2004)
15. Cannataro, M., Talia, D., Trunfio, P.: Distributed data mining on the grid. Future Generation Computer Systems 18(8), 1101–1112 (2002)
16. Cormode, G., Muthukrishnan, S., Zhuang, W.: Conquering the divide: Continuous clustering of distributed data streams. In: ICDE 2007, pp. 1036–1045 (2007)

17. Gama, J., Gaber, M.M. (eds.): Learning from Data Streams – Processing techniques in Sensor Networks. Springer, Heidelberg (2007)
18. Gaber, M.M., Yu, P.S.: A framework for resource-aware knowledge discovery in data streams: a holistic approach with its application to clustering. In: ACM Symposium Applied Computing, pp. 649–656. ACM Press, New York (2006)
19. Motwani, R., Raghavan, P.: Randomized Algorithms. Cambridge University Press, Cambridge (1997)
20. Vapnik, V.: The nature of statistical learning theory. Springer, Heidelberg (1995)
21. Manku, G.S., Motwani, R.: Approximate frequency counts over data streams. In: Proceedings of the 28th International Conference on Very Large Data Bases (2002)
22. Cormode, G., Muthukrishnan, S.: An improved data stream summary: The count-min sketch and its applications. In: Farach-Colton, M. (ed.) LATIN 2004. LNCS, vol. 2976, pp. 29–38. Springer, Heidelberg (2004)
23. Domingos, P., Hulten, G.: Mining High-Speed Data Streams. In: Parsa, I., Ramakrishnan, R., Stolfo, S. (eds.) Proceedings of the ACM Sixth International Conference on Knowledge Discovery and Data Mining, pp. 71–80. ACM Press, New York (2000)
24. Hulten, G., Domingos, P.: Catching up with the data: research issues in mining data streams. In: Proc. of Workshop on Research issues in Data Mining and Knowledge Discovery (2001)
25. Gama, J., Rocha, R., Medas, P.: Accurate decision trees for mining high-speed data streams. In: KDD 2003: Proceedings of the Ninth ACM SIGKDD International Conference on Knowledge Discovery and Data Mining, pp. 523–528. ACM, New York (2003)
26. Hulten, G., Spencer, L., Domingos, P.: Mining time-changing data streams. In: Proceedings of the 7th ACM SIGKDD International Conference on Knowledge Discovery and Data Mining, pp. 97–106. ACM Press, New York (2001)
27. Rodrigues, P.P., Gama, J., Pedroso, J.: Hierarchical clustering of time series data streams. IEEE Transactions on Knowledge and Data Engineering 20(5), 615–627 (2008)
28. Bar-Or, A., Keren, D., Schuster, A., Wolff, R.: Hierarchical decision tree induction in distributed genomic databases. IEEE Transactions on Knowledge and Data Engineering 17(8), 1138–1151 (2005)
29. Kifer, D., Ben-David, S., Gehrke, J.: Detecting change in data streams. In: VLDB 2004: Proceedings of the 30th International Conference on Very Large Data Bases, pp. 180–191. Morgan Kaufmann Publishers Inc., San Francisco (2004)
30. Cauwenberghs, G., Poggio, T.: Incremental and decremental support vector machine learning. In: Proceedings of the 13th Neural Information Processing Systems (2000)
31. Castillo, G., Gama, J.: An adaptive prequential learning framework for Bayesian network classifiers. In: Fürnkranz, J., Scheffer, T., Spiliopoulou, M. (eds.) PKDD 2006. LNCS (LNAI), vol. 4213, pp. 67–78. Springer, Heidelberg (2006)
32. Gama, J., Medas, P., Castillo, G., Rodrigues, P.: Learning with drift detection. In: Bazzan, A.L.C., Labidi, S. (eds.) SBIA 2004. LNCS (LNAI), vol. 3171, pp. 286–295. Springer, Heidelberg (2004)
33. Spinosa, E., Gama, J., Carvalho, A.: Cluster-based novel concept detection in data streams applied to intrusion detection in computer networks. In: Proceedings of the 2008 ACM Symposium on Applied Computing. ACM Press, New York (2008)
34. Barbara, D., Chen, P.: Using the fractal dimension to cluster datasets. In: Proc. of the 6th International Conference on Knowledge Discovery and Data Mining, pp. 260–264. ACM Press, New York (2000)

35. Kargupta, H., Sivakumar, K.: Existential Pleasures of Distributed Data Mining. In: Data Mining: Next Generation Challenges and Future Directions. AAAI/MIT Press (2004)
36. Aggarwal, C. (ed.): Data Streams – Models and Algorithms. Springer, Heidelberg (2007)
37. Wald, A.: Sequential Analysis. John Wiley and Sons, Inc., Chichester (1947)
38. Brain, D., Webb, G.: The need for low bias algorithms in classification learning from large data sets. In: Elomaa, T., Mannila, H., Toivonen, H. (eds.) PKDD 2002. LNCS (LNAI), vol. 2431, pp. 62–73. Springer, Heidelberg (2002)

Ubiquitous Data

Andreas Hotho[1], Rasmus Ulslev Pedersen[2], and Michael Wurst[3],[*]

[1] University Kassel
Depart. of Electrical Engineering/Computer Science
Knowledge and Data Engineering Group
[2] Copenhagen Business School,
Dept. of Informatics, Embedded Software Lab,
Copenhagen, Denmark
[3] Technical University Dortmund, Computer Science LS8
44221 Dortmund, Germany
hotho@cs.uni-kassel.de, rup.inf@cbs.dk, michael.wurst@uni-dortmund.de

Ubiquitous knowledge discovery systems must be captured from many different perspectives. In earlier chapters, aspects like machine learning, underlying network technologies etc. were described. An essential component, which we shall discuss now, is still missing: *Ubiquitous Data*. While data themselves are a central part of the knowledge discovery process, in a ubiquitous setting new challenges arise. In this context, the emergence of data itself plays a large role, therefore we label this part of *KDubiq* systems *ubiquitous data*. It clarifies the KDubiq challenges related to the multitude of available data and what we must do before we can tap into this rich information source.

First, we discuss key characteristics of ubiquitous data. Then we provide selected application cases which may seem distant at first, but after further analysis display a set of clear commonalities. The first example comes from Web 2.0 and includes network mining and social networks. Later, we look at sensor networks and wireless sensor networks in particular. These examples provide a broad view of the types of ubiquitous data that exist. They also emphasize the difficult nature of ubiquitous data from an analysis/knowledge discovery point of view, such as overlapping or contradicting data. Finally, we provide a vision how to cope with current and future challenges of ubiquitous data in KDubiq.

1 Characteristics of Ubiquitous Data

Data mining can be applied to many different types of data (cf. [1]). These data types range from simple atomic types to more complex constructs such as time series, relational data, graphs, images and many others. Over the years, many methods were developed to transform and incorporate these different data types in the overall data mining process.

One aspect that has been studied only briefly so far is ubiquity. In many current applications, data emerges in an ubiquitous way rather than in cooperated data bases. Orthogonal to the actual type of data, this adds new facets to the

---

[*] With contributions from N. Korfiatis, C. Körner, C. Lauth, C. Schmitz.

M. May and L. Saitta (Eds.): Ubiquitous Knowledge Discovery, LNAI 6202, pp. 61–74, 2010.
© Springer-Verlag Berlin Heidelberg 2010

data that is collected. We will refer to such data as *ubiquitous data* and to the corresponding augmented data types as *ubiquitous data types*.

Definition of Ubiquitous Data. We can define ubiquitous data as such data that emerges in an asynchronous, decentralized way from many different, loosely-coupled, partially overlapping, possibly contradicting sources.

A first characteristic of such ubiquitous scenarios is that data is produced asynchronously in a highly decentralized way. Collecting such data in a central data warehouse can be prohibitively expensive and even impossible in some applications (if no fast and reliable network connection is available). This is even more challenging if data emerges rather quickly. Often there is no data repository, growing over time, but rather a stream of data (see Chapter 3). Therefore, ubiquitous data measurements are usually annotated with a timestamp, as information about time is essential in the subsequent data mining process. Furthermore, most ubiquitous data has an element of location associated. This enables the so-called space-time trajectories, which have recently become a topic of significant research, e.g. in the GeoPKDD project [2].

Another aspect of ubiquitous scenarios is that usually many different data types are involved that make up a whole. For a user of the popular MySpace platform, for instance, it is very natural to deal with a combination of website, pictures, audio tracks, forums and other information. In sensor networks, different sensors may measure image data, sound, video, temperature, light, acceleration, etc. to obtain a complete picture of a situation. From a data mining point of view, this situation leads to highly complex preprocessing and data integration issues (involving, e.g., sensor fusion or ontology mapping). Furthermore, many of the data types involved in such scenarios have a complex inner structure, e.g. blog networks or cascaded sensors.

A third characteristic feature of ubiquitous data mining scenarios is the fact that data in almost all cases emerges from a very high number of partially overlapping, loosely connected sources. An example are social bookmarking systems. Users are allowed to annotate resources with arbitrary textual expressions, called tags. Users here represent independent data sources that may be in different relation to each other. Tags assigned by different users can complement each other, they can be contradictory, they can be redundant, etc. Processing data from many different sources is significantly more challenging than processing data from a single source or a small number of sources, because contradictions and overlaps cannot be resolved manually, like in data warehouse systems. A common characteristic of ubiquitous data types is thus, that each data point or measurement is additionally annotated with a data source and that this annotation plays a central role in the data mining process.

We will exemplify the implications of ubiquitous data types on two different areas: data mining in Web 2.0 environments and in distributed sensor networks.

Web 2.0 denotes a shift in the way the Internet is used. Instead of a small number of users that produces all content, now a very large number of users actively contributes to extend the internet. These contributions range from blogs, social bookmarking, videos, pictures, articles in discussion, ratings, reviews, and many others. Applying data mining to this data is, on the one hand, very appealing, as it helps to consolidate this information and to make it maximally useful. On the other hand it is very challenging as different data types are involved, data is produced by many different users distributed all over the world and is often noisy and contradicting, partially overlapping etc. While this kind of data can be stored centrally, it emerges in a distributed way. Because of its huge amount of data, consolidating and cleaning this data manually is not a reasonable option. It thus falls under the discussed ubiquitous paradigm.

Belonging to a seemingly completely different area are sensor networks. In sensor networks, a large number of distributed, partially connected sensors are used to measure some properties in a given domain. Measurements of the different sensors may overlap (neighboring sensors) or contradict (if a sensor does not work properly). Sensors can also be connected in networks of cascading sensors. Finally, sensors can capture many different data types, such as images, video, etc. A good illustration is the popular sensor network systems is TinyOS. On the website (www.tinyos.net) there are numerous examples of real applications: mobile agents, magnetic measurements, and robotic grids just to name a few.

On a second sight, sensor networks and Web 2.0 data is very similar, as both expose the typical characteristics of ubiquitous data. In both cases, data points are annotated with a source (the user or the sensor) that plays an important role in the data mining process or with some kind of timestamp to describe the history of the data. In both areas we face different data types that represent the same entities from different perspectives (e.g. MySpace profiles and different sensors in a mobile robot). Also, data emerges asynchronously in a distributed way in both areas and there is usually no way of collecting this data in a single data base.

Research in both areas has developed almost completely independently, despite these strong connections. In the following, we therefore present first challenges of ubiquitous user-centered data types with focus on Web 2.0 mining, and then research on data mining in distributed sensor networks. Finally, we discuss the common vision of both, by analyzing which challenges they have in common and which methods would be mutually applicable.

2 Web 2.0

2.1 Emerging Data from Distributed Sources

Ubiquitous data can be processed in different ways in the context of Web 2.0 applications. On the one hand, data and annotations provided by users may emerge in a distributed way and be also stored locally. Famous examples are P2P networks [3,4] or distributed agent based systems [5]. On the other hand, there are many applications in which data emerges in a distributed way, but is then collected and stored at a central server like in social bookmarking systems [6,7,8] Providing content in a Web 2.0 like way is applicable to a wide range of resources and data types, such as web pages, images, multimedia, etc. This popular principle is currently applied to many domains.

The data with which we have to deal in these applications is quite diverse. On the one hand, we have textual data, which might be partially structured (as in Wikipedia entries) or which may consist of short snippets only, such as in social bookmarking systems. Other applications, as for instance audio, image or video sharing platforms, are concerned with multimedia data.

All these applications have in common that data emerges in a ubiquitous way from many independent sources (users). This leads to similar challenges in all these applications connected to the basic characteristics of ubiquitous data, such as contradictions, overlaps and heterogeneity. Tags in social bookmarking systems may be contradicting, for instance. Also, there are many duplicate images and videos on popular multi-media sharing platforms.

The current hype around Web 2.0 applications contributes to several important challenges for future data and web mining methods. Such challenges include the analysis of loosely-coupled snippets of information, such as overlapping tag structures, homonym or synonym tags, blog networks, multimedia content, different data types etc. Other challenges arise from scalability issues or new forms of fraud and spam. Often, a more structured representation, which allows for more interaction, is needed. Mining could help here to bridge the gap between the weak knowledge representation of the Web 2.0 and the semantic web by extracting hidden patterns from the data.

In the following, we will show how the described Web 2.0 data relates with more structured data and how data mining can be applied to these data types. We will finish this section with applications showing the emergence of the ubiquitous data.

2.2   Semantic Web and Web 2.0

Web 2.0 applications often use very weak meta data representation mechanisms that reflect the distributed, loosely-coupled, nature of the underlying processes. The most basic meta data annotation is a tag, thus an arbitrary textual description that users can assign to any resource. Tags may denote any kind of information and are, by default, not structured in any way. There are several approaches to allow users to express slightly more information while tagging. Such approaches allow them to create hierarchies of tags or groups of tags [8] (sometimes referred to as aspects [10]). Tags are limited to express only very simple information. However, the same principle allowing users to create their own conceptualization in a decentralized way, is also applied to more complex data. Structured tag representations, such as XML microformats, created by users for various special applications, are a good example for this kind of knowledge representation.

All of these efforts can be denoted as local. They represent the requirements and views of individual users or of user groups. An extreme case of locality would be a user that applies tags that are private in the sense that they are not shared by any other user. Local approaches complement global approaches, such as standardized XML formats or the Semantic Web, in a way that is better applicable in highly distributed scenarios.

The Semantic Web is based on a vision of Tim Berners-Lee, the inventor of the WWW. The great success of the current WWW leads to a new challenge: a huge amount of data is interpretable by humans only; machine support is limited. Berners-Lee suggests to enrich the Web by machine-processable information which supports the user in his tasks. For instance, today's search engines are already quite powerful, but still too often return excessively large or inadequate lists of hits. Machine-processable information can point the search engine to the relevant pages and can thus improve both, precision and recall. In such global approaches, all users must obey a common scheme. For instance, authors of scientific publications usually have to classify their work according to a static classification scheme of topics provided by the publisher.

A key approach is to use data mining techniques that mediate between the different levels of meta data [10] and to use semantic web technology to store and transport the knowledge. This can be achieved, for instance, by identifying patterns such as combinations of tags that co-occur often [11], by analyzing, identifying and using the emergence structure and inherent semantics of web 2.0 systems [12,13,14] or by extracting information from natural language texts by information extraction. Then a strong knowledge representation will represent the data appropriately, can help to transfer distributed learned knowledge and act in this way as a mediator between any kind of learning system.

2.3   Peer-to-Peer Based Web 2.0 Applications

Despite of its conceptual and architectural flaws, many current Web 2.0 applications store their data centrally. This is not necessarily so. There are several approaches to combine Web 2.0 tagging or Semantic Web (SW) with P2P technology. General characteristics of P2P systems and their relevance for ubiquitous knowledge discovery have been discussed in Chapter 2, Section 1. In this section we focus on the relation between P2P, Web 2.0 and the Semantic Web.

In [3] P2P is defined in the following way:

> "Peer-to-peer is a class of applications that takes advantage of resources—storage, cycles, content, human presence—available on the edge of the Internet."

This definition combines two aspects. First, data is created by users, which act as a resource by annotating data, locations, etc. On the other hand, these users are often located at the edge of the internet, communicating only with mobile phones or handheld devices. This fully distributed nature of not only data emergence but also data storage has a strong influence on the way data is represented and processed.

The combination of Semantic Web and P2P technology, for instance, opens up a feasible way of combining rich knowledge representation formalisms, on the one hand, with low overhead but with immediate benefit, on the other hand. Another possible application of semantic P2P networks, that has been researched in projects such as Nepomuk[1], is the *Social Semantic Desktop*. Every user of desktop applications is faced with the problem that the office documents, emails, bookmarks, contacts, and many other pieces of information on their computers are being processed by isolated applications which do not maintain the semantic connections and metadata that apply to the resources. Semantic desktop applications offer approaches to deal with this problem of distributed storage based on semantic integration.

Connecting semantic desktops by a P2P network is an idea put forward by [15]. The result is called *Networked Semantic Desktop* or *Social Semantic Desktop* (the latter name being the one most commonly used today). This idea has received a lot of attention lately and has spawned a successful series of workshops [16,17]. Data mining is a very natural extension to make semantic desktop systems more flexible an adaptive. The corresponding methods must, just as for text mining, respect the local context in which the information emerged. They require, therefore, ubiquitous data mining methods almost by definition.

An example of an approach that combines social bookmarking systems and P2P technology is the Nemoz system [10]. Nemoz (see Chapter 1, Section 6.2) is a distributed media organization framework that focuses on the application of data mining in P2P networks.

Media collections are usually inherently distributed. Multimedia data is stored and managed on a large variety of different devices with very different capabilities concerning network connection and computational power. Such devices range from high performance work stations to cell phones. This demands for sophisticated methods to distribute the work load dynamically including the preparation of the data.

The key point in processing multi media data is an adequate data representation. Finding and extracting adequate features is computationally very demanding and therefore not simply applicable in an interactive end user system. The key to solve this problem is twofold. First, by distributing the work load among nodes with different computational capabilities, a much higher efficiency can be achieved, which allows the application of data mining even on devices with only little resources. Second, it is possible to exploit the fact that although all local tag structures created by the users may differ in any possible way, it can be assumed that at least some of them resemble each other to some extent, e.g. many users arrange their music according to genre. Thus by collaborating in the Data Mining process, each learner can profit from the work done by other nodes [18].

Network Mining. Mining networks is of particular interest to ubiquitous data mining, as most data addressed in this book forms some kind of graph or network. One research area, which has addressed the analysis of graphs comparable to the one discussed in this chapter, is the area of "Social Network Analysis". Therefore, methods developed in that area are a good starting point to develop and analyze

---

[1] http://nepomuk.semanticdesktop.org/

other algorithms for mining ubiquitous data. We will shortly recall the ideas behind (social) network mining.

A (social) network (G) is a graph that consists of a set of nodes (V) and a set of edges (E) such as: $G := (V, E)$ where V is the set of network actors (e.g. people or institutions) and E is the set of relations between them. Most Web 2.0 systems can be represented as such a graph. Social networks can be explicit, such as in popular applications as LinkIn. They often, however, emerge rather implicitly by users communicating or collaborating via mail, instant messaging or mobile phones. This distributed emergence of social networks through communication patterns is an important challenge for ubiquitous data mining methods. Methods developed in this area are especially important as they provide valuable insights into the relationship among Web 2.0 participants.

Social network mining tackles the extraction of social network data from sources that contain recordings of interactions between social entities (e.g. emails). Social network data describes structural relations between social entities, such as blog authors or public profiles of users. Social network data consists of two types of variables: structural and compositional. *Compositional variables* represent the attributes of the entities that form the social network. That kind of data can be, for instance, the topic of the webpage (e.g. portal, homepage) or things related to the profile of the social entity they describe (e.g. age, address, profession, etc). *Structural variables* define the ties between the network entities (E) and the mode under which the network is formed. The term mode refers to the number of entities which the structural variables address in the network.

Current social network mining methods process data usually in a centralized fashion. This will soon no longer be possible. With the emergence of massive amounts of implicit social networking data, such as phone data or email data, the analysis of this data can not be performed in a centralized way anymore. Rather, it will be important to find highly distributed approaches such that communities or other strongly connected components in the social graph are identified locally, bottom-up, instead of doing this centrally, top-down. Research on such methods will certainly play an essential role for the scalability of future social networking systems.

3   Sensor Networks

Today, only few people can participate in everyday life without leaving traces of their actions. The mass usage of mobile phones, GPS traveling assistance and the spreading application of radio frequency identification (RFID) technology to pursue the lifetime of goods are only the most prominent examples of how collections of data appear in everyday life. In addition, the costs of sensor nodes are constantly decreasing, which makes their future application very attractive and provides for many large-scale sources of ubiquitous data. However, the knowledge extraction from ubiquitous data sources and careful handling of private data are only in their infancy (for privacy, see the discussion in Chapter 5).

Ubiquitous systems come in different forms as well. Some ubiquitous systems are web-based, others are rooted in the physical surroundings. Sensor networks belong to the latter category. In the following, we will introduce sensor networks and discuss selected aspects related to the ubiquitous knowledge discovery process. At all times these systems are subject to severe resource constraints (see Chapter 3, Section 2. Our discussion will focus on data aspects. Distributed computing aspects of wireless sensor networks have been discussed in Chapter 2, Section 3.

3.1   Sensors as Data Sources

Just as humans possess senses, there are many different kinds of sensors to form a sensor network. Basic sensors record temperature, light or sound and form fundamental parts in today's industrial applications. New types of sensing emerged in the areas of health services and environmental monitoring. Think, for example, of nearly invisible devices to monitor heart beats or to measure diverse chemical substances, acid levels or carbon dioxide in natural surroundings. The data sources record phenomenons that spread in space. It is also possible to record characteristics of the landscape itself as in satellite recordings and to monitor the spatial location of objects. The latter has become very popular with the emergence of mobile phones, GPS and RFID.

Sensors. There has been significant work by IEEE to provide information regarding the output of sensors. This is a pre-requisite for getting access and using the information in an ubiquitous knowledge discovery system. Sensors with these capabilities can be referred to as smart transducers. This is also described in the IEEE 1451 (proposed) standards.

Sensors may provide information according to the specifications in the IEEE 1451 data sheets. The electrical connections of this system are defined for 2, 3, or 4 wires. It works by a template that provides semantic information regarding the data from the sensor.

3.2   Sensor Network Modeling Languages

In wireless sensor networks (see Chapter 2, Section 3) there has been some work on a sensor model language called sensorML. It uses XML to encode and describe sensor models. The whole process ranging from input to output and associated parameters can be described using this language. It is part of the standards proposed by the Open Geospatial Consortium.

SensorML. SensorML is a XML schema-based markup language to describe sensor systems. This language can describe both stationary and dynamic sensors such as robotic sensor systems. Furthermore, it can describe sensors placed inside the monitored object as well as sensors placed remote with respect to the object. Processes describe which input they measure, which output they produce. In addition the method used to produce the output can be described. One example could be a 2-dimensional input sampled at 1 Hz to produce one classification output with a binary support vector machine. In wireless sensor networks, data types are often real values, representing specific physical phenomena. This can include GPS sensors, mobile heart monitors, web cams, satellite-borne earth images. It is already clear from this wide range of wireless sensor networks data that any ubiquitous knowledge discovery agent needs meta information regarding the sensor data. It can be challenging to provide this information in situ. Therefore, an external description of a sensor and the services it can provide is necessary. SensorML is one such approach.

3.3   Wireless Sensor Network Applications

TinyOS is a popular sensor network platform. Other sensor network platforms are starting to emerge such as the Java-based Sun SPOT from Sun Microsystems. The list of example applications for wireless sensor networks is long. However, data management in these distributed loosely coupled systems are difficult. Later in this section we will discuss TinyDB [19] as example abstraction of this problem. Furthermore, a company named Sentilla has taken an interesting approach to Java object serialization which we shall also touch upon. Next we look at some general examples which are all related to ubiquitous data in some way.

We find examples like mobile agents, magnetic measurements, robotic grids, pre- and in-hospital monitoring, volcanic eruption detection, hog monitoring, distributed location determination, structural health monitoring, rescue equipment, neural wireless interface, oil/gas exploitation, and elite athlete monitoring. There are some performance challenges described in Chapter 7, Section 7, which add a systems perspective to some of these applications.

Recently, a new experimental platform for sensor networks using LEGO MINDSTORMS NXT has been introduced by Pedersen [20]. It runs the small wireless operating system TinyOS, while providing access to many different sensors on this popular educational platform.

The traditional message abstraction in TinyOS is a message structure, which is similar to a `struct` in the C programming language. This approach is fine for the application domain because the individual sensors know the context of the data. Now we can look at two more general approaches.

In the context of TinyOS, which was just introduced, there is an data abstraction called TinyDB. In TinyDB, the data in the distributed sensor network are stored within the network nodes (called motes in TinyOS terms). What is interesting is that the data can be retrieved using SQL language which is familiar to many people. That is one example from the TinyOS area. Java-based sensor networks are also emerging and a company like Sentilla presented their approach to data management at the JavaOne 2008 conference. They acknowledge that the packet payload in a 802.15.4 network is so small that traditional Java object serialization is not feasible. Instead they send raw packets of data and let a gateway mote server translated the raw data to Java objects again. In that way the data can be used like traditional Java objects in the client application. To summarize, we have shown TinyDB as one SQL-based abstraction of ubiquitous data and we have shown another abstraction of ubiquitous data in terms of Java objects. Common to both solutions are that they abstract the access to raw ubiquitous data.

3.4   Object Monitoring in Space and Time

Trajectory data will become a highly important data type in the near future. An example for the monitoring in space and time using GPS trajectories has been given in Chapter 1, Section 6.1, in the activity recognition example. A trajectory is the path of a moving object within time and space. It can be viewed as a function mapping each moment in time to the location of a given object. In practice, trajectories are represented as (finite) number of pairs $(time, location)$ where both time and space have been discretized [21]. Depending on the measurement technology, recordings occur at regular or irregular time intervals. The location can either be expressed within a coordinate system, e.g. latitude / longitude, or symbolic coordinates, as for example cell identifiers of mobile phone data.

GPS Data. GPS technology records the x,y-position of an object every second and thus provides a quite accurate picture about the object's position and motion. If GPS signals are used without other assisting technologies, a standard error of 10 - 25 m or 5 - 10 m for newer devices is made during location approximation. However, GPS signals are blocked by buildings, and measurements indoors, inside of tunnels or in alleys between tall buildings may be lost.

Mobile Phone Data. In cellular networks, mobile phones perform a location update procedure in regular or irregular intervals to communicate their current location area. In addition, cell identifiers are recorded during the time of a call. It is therefore possible to derive movement information or call densities from logging data of mobile phone networks. The data, however, possesses a coarse level of granularity. Cell sizes of GSM networks range between 0.1 and 1 km in the city centre but can extend as far as 30 km in rural areas. An additional uncertainty is introduced by overlapping cells. Depending on the signal strength and workload of a base station antenna, a handover to another cell may occur although the user does not change his or her position.

Both sources for trajectory data have their strengths and weaknesses. While GPS trajectories are very accurate in location, they may contain gaps and the number of available trajectories is usually small. In comparison, mobile phone data have a low resolution but are produced in great masses. One challenge lies in the combination of both data sources and mutual exploitation of advantages.

4   Common Characteristics of Data from Web 2.0 and Sensor Networks

In both areas, Web 2.0 environments and sensor networks, we face ubiquitous data and ubiquitous data types. Ubiquitous data types augment traditional data types like texts, photos, videos, bookmarks (in the case of Web 2.0) or data types like temperature, sound and light (in the case of sensor networks) with several novel aspects: a source identifier (user id or sensor id), a timestamp (either logical or real time) possibly a location or more specialized features depending on the scenario. On the one hand, this information is very valuable and provides new possibilities for the application of data mining. On the other hand, it leads to a set of challenges in both areas. The massive amount of different sources and the necessity to preprocess data locally, "in context" goes beyond the capabilities of current data mining systems. In the following, we will address the common characteristics of both application areas and involved data types.

A first important common challenge is the large number of heterogeneous sources in both application areas. In Web 2.0 applications, we have a large user base, providing different kinds of meta data, that partially overlap, contradict or complement each other. In sensor networks, we have a number of sensors (sometimes referred to as smart dust) that measure partially overlapping entities. Sensor data maybe contradictory as well, e.g., if a sensor is faulty. In both cases, there is a need to aggregate measurements and annotations into a smaller, concise set of information. For Web 2.0 applications, this can be approached by applying advanced clustering methods. In sensor networks, the data are combined using sensor fusion. There is an additional challenge as the data streams cannot be cached on the sensor nodes, so the sensor fusion has to be done in real time. Existing methods for both, tag aggregation and sensor fusion, are not fully satisfying. For instance, many graph mining methods are not directly applicable to the hypergraph structure of folksonomies [14] that emerge from the augmentation of annotations with a user id or a timestamp. The same holds for the multi-mode structure of blog networks and wiki page editing networks. Common to both scenarios is that the distributed computers/sensors try to collaborate. It could have been the other way around such that each computing unit needed to be aware of malicious information. In this sense we have a *cooperative* scenario.

Another problem common to data that emerges in a loosely coupled, distributed way is that object references are hard to define globally. One solution for the web is to use the URI as identifier. For sensor networks, RFID provides a possible solution. Both solutions are not able to cope with all challenges that arise from this, such as duplication detection problems, missing and wrong links and information. This topic will require substantial research in the next years.

In both areas, sources are often structured. In blog networks, individual blogs can be mash-ups of other blogs. In sensor networks, sensor nodes are often explicitly associated. This association is a sort of meta tagging of the data. Often the nodes can be localized with respect to one another. Using many sensors results in higher reliability, but also in redundancy. There are currently few methods that take the interrelation of data sources into account.

In both areas, problems can only be tackled by applying highly distributed processing and intelligence. In Web 2.0 applications, these are intelligent clients performing all kinds of support for the user (mostly Ajax based). In sensor networks, distributed units performs some kind of autonomous computing to deal with the large number of measurements under restricted resources. The resource constraints can be addressed with local computing that can save on the radio use. This can extend the deployment of the sensor network.

Data and meta data is of essential importance for knowledge discovery. A study of ubiquitous data in the two area of interest reveals many common characteristics and challenges. In our opinion, several methods of distributed processing and intelligence are equally applicable in both areas. In particular, with an increasing amount of data that emerges in social applications, there is an increasing need to process this data locally. Several methods, as highly distributed sensor fusion or aggregation, could be applicable in this domain. This would enable future Web 2.0 applications to process data in place and in context, increasing their scalability and the soundness of results.

To achieve the benefits of such synergy, researchers in both areas have to collaborate closely to make ubiquitous knowledge discovery an everyday experience.

5   Emerging Challenges and Future Issues

There are a lot of open issues which need to be solved to make different kinds of ubiquitous data useful by applying knowledge discovery algorithms to it. In this chapter we have discussed two example domains: data from Web 2.0 applications and sensor networks. There are several issues still open. First, there is the need for a coherent implementation-independent description of ubiquitous data including more semantics. Second, there is a need to describe the dynamic nature of ubiquitous data and its refinement over time. Third, there is a need for patterns and best practices of how to implement ubiquitous data descriptors in resource constrained environments.

In terms of data description, we find that there are several existing approaches to these problems in both Web 2.0 and in sensor networks. These are the two dominant examples we have used in this chapter on ubiquitous data. To make the data available for various data mining and knowledge discovery tasks we need to first describe the data. One step to get an independent description is to make use of semantic web technology and combine this with the advantages of Web 2.0. This is often referred to as Web 3.0. Semantic web approaches provide also nice solutions to describe sensor data in an abstract way. A more abstract and machine readable description of data allows for a better use of data. But to make this vision reality in all ubiquitous environments, one has to solve problems resulting from limited resources, limited bandwidth or missing and contradicting data.

The second dimension of ubiquitous data description is the dynamic nature of the data and the need to describe updates to the data. Today, most of the data descriptions (like XML for example) is targeted toward static data. However, data changes rapidly in ubiquitous environments, and furthermore we may use techniques like machine learning on streaming data to refine the information content of the data on the fly. This dynamic nature of the system is not reflected in current approaches.

The third part of the vision is to provide practical implementations. We discussed earlier in this chapter two different solutions in sensor networks for an SQL abstraction and a Java abstraction which were mapped to the constraints of the sensor network. As of today, there is not any standard solution to this problem. Semantic Web technology provides a principle solution to store such data in a reusable way but does not provide implementations respecting the resources constraints of ubiquitous devices.

We need a systematic way to design our data representation in terms of constraints such as system related constraints (CPU, memory, battery limitations etc.), data mining related constraints (privacy, dynamic data, etc.), and other constraints (domain specific such as data quality constraints in social network mining).

With these three main dimensions of ubiquitous data representation, we can look at the next steps in the development and application of knowledge discovery algorithm to ubiquitous data. These steps are centered around the combination of Web 2.0, semantic web, and the physical world of sensors including all restrictions of small devices which can be seen as the next step of the evolution of Web. As this will be, however, a long way to go, we may label this vision *Web 4.0*.

References

1. Hand, D., Mannila, H., Smyth, P.: Principles of Data Mining. MIT Press, Cambridge (2001)
2. Giannotti, F., Pedreschi, D. (eds.): Mobility, privacy, and geography: a knowledge discovery perspective. Springer, Heidelberg (2008)
3. Shirky, C.: Listening to Napster. In: [4], pp. 21–37
4. Oram, A. (ed.): Peer-to-Peer. O'Reilly, Sebastopol (2001)
5. Weiß, G. (ed.): Multiagent Systems: A Modern Approach to Distributed Artificial Intelligence. MIT Press, Cambridge (1999)
6. Golder, S., Huberman, B.A.: The structure of collaborative tagging systems (2005)
7. Hammond, T., Hannay, T., Lund, B., Scott, J.: Social Bookmarking Tools (I). D-Lib Magazine (2005)
8. Jäschke, R., Hotho, A., Schmitz, C., Stumme, G.: Analysis of the publication sharing behaviour in BibSonomy. In: Priss, U., Polovina, S., Hill, R. (eds.) ICCS 2007. LNCS (LNAI), vol. 4604, pp. 283–295. Springer, Heidelberg (2007)
9. Flasch, O., Kaspari, A., Morik, K., Wurst, M.: Aspect-based tagging for collaborative media organisation. In: Proceedings of the ECML/PKDD Workshop on Ubiquitous Knowledge Discovery for Users (2006)
10. Stumme, G., Hotho, A., Berendt, B.: Semantic web mining - state of the art and future directions. Journal of Web Semantics 4(2), 124–143 (2006)
11. Schmitz, C., Hotho, A., Jschke, R., Stumme, G.: Mining association rules in folksonomies. In: Batagelj, V., Bock, H.H., Ferligoj, A., Ziberna, A. (eds.) Data Science and Classification. Proceedings of the 10th IFCS Conf. Studies in Classification, Data Analysis and Knowledge Organization, pp. 261–270. Springer, Heidelberg (2006)
12. Mika, P.: Ontologies are us: A unified model of social networks and semantics. In: Gil, Y., Motta, E., Benjamins, V.R., Musen, M.A. (eds.) ISWC 2005. LNCS, vol. 3729, pp. 522–536. Springer, Heidelberg (2005)

13. Hotho, A., Jäschke, R., Schmitz, C., Stumme, G.: Information retrieval in folksonomies: Search and ranking. In: Sure, Y., Domingue, J. (eds.) ESWC 2006. LNCS, vol. 4011, pp. 411–426. Springer, Heidelberg (2006)
14. Cattuto, C., Schmitz, C., Baldassarri, A., Servedio, V.D.P., Loreto, V., Hotho, A., Grahl, M., Stumme, G.: Network properties of folksonomies. AI Communications 20(4), 245–262 (2007)
15. Decker, S., Frank, M.R.: The Networked Semantic Desktop. In: Proc. WWW Workshop on Application Design, Development and Implementation Issues in the Semantic Web, New York (2004)
16. Decker, S., Park, J., Quan, D., Sauermann, L. (eds.): The Semantic Desktop - Next Generation Information Management & Collaboration Infrastructure. Proc. of Semantic Desktop Workshop at the ISWC 2005, CEUR Workshop Proceedings, vol. 175 (2005), ISSN: 1613–0073
17. Decker, S., Park, J., Sauermann, L., Auer, S., Handschuh, S. (eds.): Proceedings of the Semantic Desktop and Social Semantic Collaboration Workshop (SemDesk 2006) at the ISWC 2006, Proceedings of the Semantic Desktop and Social Semantic Collaboration Workshop (SemDesk 2006) at the ISWC 2006. CEUR-WS, vol. 202 (2006)
18. Wurst, M., Morik, K.: Distributed feature extraction in a p2p setting - a case study. Future Generation Computer Systems, Special Issue on Data Mining (2006)
19. Madden, S.R., Franklin, M.J., Hellerstein, J.M., Hong, W.: Tinydb: an acquisitional query processing system for sensor networks. ACM Trans. Database Syst. 30(1), 122–173 (2005)
20. Pedersen, R.U.: Tinyos education with lego mindstorms nxt. In: Gama, J., Gaber, M.M. (eds.) Learning from Data Streams. Processing Techniques in Sensor Networks, pp. 231–241. Springer, Heidelberg (2007)
21. Andrienko, N., Andrienko, A., Pelekis, N., Spaccapietra, S.: Basic concepts of movement data. In: Mobility, Privacy and Geography: a Knowledge Discovery Perspective. Springer, Heidelberg (2008)

Privacy and Security in Ubiquitous Knowledge Discovery

Fosca Giannotti[1] and Yücel Saygin[2,⋆]

[1] KDD Laboratory,
ISTI-CNR,
Via G. Moruzzi 1,
56124 PISA, Italy
[2] Faculty of Engineering and Natural Sciences
Sabanci University,
34965, Istanbul, Turkey
fosca.giannotti@isti.cnr.it, ysaygin@sabanciuniv.edu

When we consider wearable computers, smart homes, and various sensors omnipresent in our environment and the data collected by such systems, we can see that ensuring privacy and security is a big challenge in ubiquitous computing (UbiComp). This is due to the fact that ubiquitous computing is highly integrated into our daily lives, making the dependability of these systems ever more central and yet difficult to control.

By dependability we mean safety, security and privacy in these systems (see Figure 1). The safety of a system is concerned with protecting the environment of the system against harm to human lives as well as major monetary loss. The security of a system is concerned with the confidentiality, integrity and availability, whereas the privacy of a system is concerned with the protection of personal or other private information belonging to stakeholders and processed by the system.

The information collected and revealed through UbiComp environments has effects on the privacy of individuals, as well as institutions and organizations. In order to utilize the advantages and minimize vulnerabilities, we must address security and privacy issues in UbiComp environments. Security in UbiComp refers to two things. First of all, security is necessary to guarantee safety and to protect these systems against threats from malicious attackers in such widely distributed open systems. Also, security refers to the necessary analysis and mechanisms to preserve the privacy of the multiple stakeholders interacting with UbiComp environments.

UbiComp also introduces new security and privacy risks, since wireless ad hoc networking will be widely used to interconnect devices with limited protection mechanisms collecting vast amounts of data. We may not even be aware of the type, volume, and resolution of the collected data. Also, in Ubicomp attacks will also become omnipresent, and the effects of security breaches are expected to be

---

⋆ With contributions from M. Atzori, F. Bonchi, M. Damiani, A. Friedman, S. Gurses, D. Malerba, D. Pedreschi, T. Santen.

M. May and L. Saitta (Eds.): Ubiquitous Knowledge Discovery, LNAI 6202, pp. 75–89, 2010.

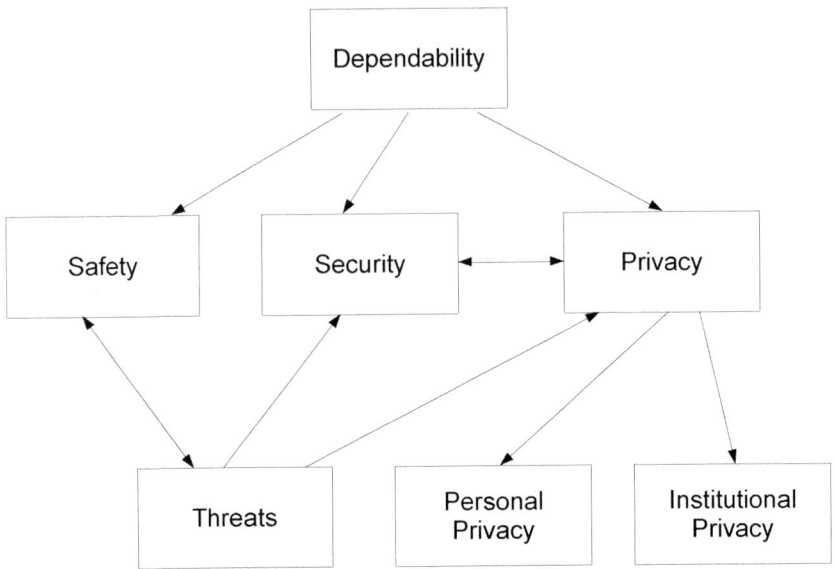

Fig. 1. Dependability of ubiquitous systems

more visible. Therefore, addressing the privacy and security issues in ubiquitous environments is critical for its deployment and social acceptance.

Privacy and security in KDubiq has two aspects: (1) The new challenges brought up by KDubiq in terms of privacy and security. (2) How can KDubiq improve the privacy and security in ubiquitous computing environments? Application of knowledge discovery to improve security of systems has been studied for a while but the investigation of privacy in knowledge discovery is a new field of research. In this chapter, we provide a brief state of the art in privacy and security issues in ubiquitous computing, then privacy and security aspects of data mining, together with a roadmap toward privacy and security aspects of ubiquitous knowledge discovery that could guide the researchers who want to work in the area.

The following properties of ubiquitous computing environments will be the key to our discussions since they combine different multidisciplinary research:

– Data is highly distributed
– Data may be online/streaming as well as historical
– Mobility: space and time components are added to the data

All of the above issues are being investigated in isolation by researchers. In case of ubiquitous knowledge discovery, we need to have an integrated approach to combine these multidisciplinary research results.

1   Basic Concepts of Security and Privacy

Security and privacy can be thought of as two related and complementary issues in KDubiq. Actually, heated debates on how to draw a line between privacy and security research, how they are related and how they are different are still ongoing. While data and network security are indispensable, this is not enough by itself to achieve security and privacy in KDubiq. The protection of our data against unauthorized access or tampering is important. Yet, we need to go beyond and make sure that those who have access to the increasing amounts of data in ubiquitous environments also do not misuse it. For that purpose, we need to keep track of who is accessing what, for what purpose, and for how long.

In classical security, the best way to keep information confidential is to avoid retention of data related to that information. This also holds true for knowledge discovery. In order to minimize the risk of compromising privacy, the purpose for which data is to be collected and processed needs to be made explicit. On this basis, the collection of data that does not contribute to the purpose of the knowledge discovery task can be avoided. In ubiquitous systems, avoiding retention of unnecessary data also has an added benefit in system performance, since it reduces the workload of the (computationally weak) devices planned for use in those systems.

1.1   Security Concepts

Security in networked computers is still an ongoing research area. The goals are to ensure availability, integrity, authenticity, and confidentiality. In the case of ubiquitous computing, constraints such as network bandwidth, battery life, and unreliable networking make things even more complicated.

Confidentiality. One of the main security issues is confidentiality, which means to keep information confidential from unauthorized principals. According to the network security terminology, the "principal" is the entity in a computer system to which authorizations are granted [1]. Wireless networking used in ubiquitous computing is structurally more vulnerable to eavesdropping attacks since information is radiated to anyone within range. In standard distributed networks, the main mechanism to protect the confidentiality of a message sent from a sender to a receiver is encryption together with principal identification and authentication. Encryption makes it easy to protect the content of the message;However, the time, the source, and the destination of the communication, as well as the fact that a conversation is taking place, remain observable and the ubiquitous computing infrastructure can become a tool for ubiquitous surveillance of unaware victims [2].

Authenticity. Authenticity is meant to assert the identity of the principals taking part in the computer system. Since in many cases the privileges of a principal and the kind of data to which it is exposed are determined based on its identity, prevention of impersonation is a key concept in ensuring the security of the system. One of the established ways to achieve this goal is by use of "digital signatures" [3].

Integrity. Integrity is the property that data has not been modified in unauthorized ways. Since it is impossible to prevent an attacker who has control of the transmission channel from altering the message, integrity preservation should be intended as "ensuring that nobody can alter a message without the receiver noticing it" [4]. In ubiquitous computing, integrity also concerns the authenticity of the collected data with respect to the state of the real world. "Perceptive attacks", that is, the misuse of the UbiComp system's augmented capabilities to provide false/inaccurate environmental context, are a kind of integrity attacks specific to UbiComp applications [5]. Indeed, the environmental context has proven to be an important contribution to UbiComp applications; the decision whether heating has to be automatically switched on or off in an office may depend not only on the temperature in the room, but also on the presence of people in the office. Nevertheless, context generally seems to be the root vulnerability in UbiComp systems [6].

Availability. Finally, availability is the property of a system that always and in a timely way honors any legitimate request by authorized principals. While confidentiality and integrity are essentially concerned with what a user is allowed to do, availability is concerned with what an authorized user is actually able to do. The violation the availability property results in a "denial of service" ("DoS" attack), typically obtained by using up all available resources. DoS attacks constitute one of the major current security problems, especially in UbiComp environments that are based on networked systems.

Some "Interactive attacks", that is, the misuse of the UbiComp system's augmented facilities for responding to or receiving environmental signals, can be considered as a form of availability attacks specific of UbiComp applications. Indeed, when an interactive attack occurs, the UbiComp system might simply not react properly to some environmental signals because it is too busy processing other requests or because its reaction does not reach the environmental context. For instance, a wireless-equipped car whose processor is flooded with requests from a malicious node of an ad hoc network, will not be able to timely react to environmental signals that indicate a hazardous situation.

1.2 Privacy Concepts

Research on privacy enhanced technology is a young field which is gaining momentum. There are currently different proposals for terminology on privacy [7,8]. In the following, we use definition given in [7].

Anonymity. Anonymity ensures that a user may use a resource or service without disclosing the user's identity. Notice that this definition is quite different from the one described in [8], where anonymity is the state of being not identifiable within a set of subjects, the so-called anonymity set. The anonymity set is the set of all possible subjects. With respect to acting entities, the anonymity set consists of the subjects who might cause an action. With respect to addressees, the anonymity set consists of the subjects who might be addressed. Therefore, a sender may be anonymous only within a set of potential senders, his/her sender anonymity set, which itself may be a subset of all subjects worldwide who may send messages from time to time. The same is true for the recipient, who may be anonymous within a set of potential recipients, which form his/her recipient anonymity set. Both anonymity sets may be disjoint, be the same, or they may overlap. The anonymity sets may vary over time. The bigger the anonymity set is, the better in terms of individual privacy.

Pseudonymity. According to [7], pseudonymity ensures that a user may use a resource or service without disclosing his or her user identity, but can still be accountable for that use. With pseudonymity, real identities are substituted with pseudonyms (pseudo-identifier), i.e., identifiers that the attacker cannot link to a specific individual. The main difference with respect to anonymity is that actors are "not" indistinguishable, since the attacker can recognize them through a pseudo-identifier, but he cannot link them to real identities.

Unlinkability. Unlinkability ensures that a user may make multiple uses of resources or services without others being able to link these uses together. In general, the idea is that the probability of interesting items being related from the attacker's perspective stays the same before (a-priori knowledge) and after the attacker's observation (a-posteriori knowledge). Therefore, the system must not release any information that improves the attacker's knowledge on links between items of interest.

Unobservability. Unobservability ensures that a user may use a resource or service without others, especially third parties, being able to observe that the resource or service is being used. The focus here is not on the identity of individuals, but rather on the items of interests, that cannot be observed by the attacker. Providing no information about resources clearly increments privacy protection of who is accessing them.

2 State of the Art

Security is an established research field. It covers areas such as network security, data security, computer systems security, intrusion detection and prevention, and computer forensics. Access control is particularly important in data security. It contributes to the integrity and confidentiality of processed data. It also contributes to computer systems security because it controls who can use which services of a system. Intrusion detection and prevention address the problem of bypassing the access control mechanism of a computer system.

Privacy can be considered to be a social issue. In general, privacy is the fundamental right for people to be left alone. The reader may refer to Chapter 6 for a further analysis of the concept. Ubiquitous computing may aggravate privacy concerns since it may allow others to be involved in our lives very easily.

In the following subsections, we are going to give a very brief state of the art in privacy and security issues in data mining in general, then privacy and security issues in ubiquitous computing. This will be the basis of our discussions later on.

When knowledge discovery was studied as an emerging field of research with vast applications from customer relationship management to bioinformatics, it was also considered as a tool for improving the system security. Knowledge discovery is employed for intrusion detection by modeling the normal and abnormal (or malicious) behavior in a system.

The European Commission has already shown some interest in this topic by funding the 12-months FET project CODMINE "Confidentiality of Data Against Data Mining Methods", a large scale project, GeoPKDD, supported by the Future and Emerging Technologies program of EU FP6, and the MODAP coordination action in FP7.[1] The interest of the European research community is also witnessed by some recent initiatives, such as a Workshop series on Privacy and Security Issues in Data Mining,[2] or publications [9].

2.1   Security in Ubiquitous Knowledge Discovery

When we think of knowledge discovery in ubiquitous environments, we need to take into account that the data is distributed, it is heterogeneous, and the data sources may be mobile, i.e., data is spatiotemporal.

As mentioned earlier, traffic analysis allows ubiquitous surveillance of unaware victims. Defending against monitoring network traffic between two wireless clients is a difficult problem and an active research area [10]. In addition to placing a technical barrier, to achieve a secure system, security must be integrated into software applications that make use of the wireless network. It is a challenging research problem to determine the minimal requirements to which nodes of a network must comply with, in order to prevent monitoring through robust data mining tools. Experience gained both by official statisticians who work on microdata confidentiality and by experts of inference control in statistical databases (See Microdata Confidentiality References[3]) can be a starting point to investigate the reverse problem of "data-mining-safe" systems.

A few standards have come out to alleviate the integrity problem by means of digital signatures, however they cannot solve the problem at the application level. This issue is currently emerging in the context of Web Services [11] as well as in Ubiquitous Computing. Indeed in many UbiComp applications, the corruption of a single message has no dramatic effect; what really matters is the global effect of a sequence of corrupted messages. In this case it is important

---

[1] http://dke.cti.gr/CODMINE/; http://www.geopkdd.eu/; www.modap.org

[2] Co-located with the European Conference on Principles and Practice of Knowledge Discovery in Databases, e.g.,
http://people.sabanciuniv.edu/ỹsaygin/psdm/index.html

[3] http://www.census.gov/srd/sdc/confid_ref1.htm

to develop trust policies. In human interaction, trust is often based on behavioral expectations that allow people to manage the uncertainty or risk associated with their interactions. Therefore, trust should be considered as a dynamic position, that can be formed and can evolve in time. The formation and evolution of trust, which are central to human intuition of the phenomenon, are neglected in current UbiComp systems. Certificate revocation works in relatively simple situations, while, more in general, it is important to develop the capability of choosing between alternative collaborators. Applying data mining (DM for short) to develop validation rules according to an explicit model of trust is just one of the novel potential contributions of KDubiq to UbiComp solutions. This application might be well based on the results of the IST-2001-32486 EU project SECURE "Secure Environments for Collaboration among Ubiquitous Roaming Entities".

In context-aware applications, DM can be used to capture security-relevant context of the environment in which service requests are made. In UbiComp environments, where access control procedures have a contextual element, information captured by DM methods can also be relevant for "'identification"' and/or "'authentication"' procedures [12].

The measures taken for protection against Denial of Service threats include the use of anti-virus software, firewalls and intrusion detection systems (IDS). However, while the first two solutions cannot be applied to several UbiComp applications (current wireless firewalls are not appropriate for ad hoc networks), IDS can help to monitor and analyze the traffic in UbiComp applications based on sensor networks. The analysis of audit data for the construction of intrusion detection models can be based on data mining algorithms which return frequent activity patterns [13]. However, sensor networks pose new challenging requirements to DM methods, since the solution should be fully distributed and inexpensive in terms of communication, energy, and memory requirements. In addition, the networked nature of sensor networks allows new, automated defenses against DoS attacks [14].

2.2   Privacy in Ubiquitous Knowledge Discovery

The awareness that privacy protection in knowledge discovery and data mining is a crucial issue has captured the attention of many researchers and administrators across a large number of application domains. Consequently, privacy-preserving data mining (PPDM) [15,16,17,9], i.e., the relationship of knowledge discovery to privacy, has rapidly become a hot and lively research area. This is made evident by the fact that major companies, including IBM, Microsoft, and Yahoo, are allocating significant resources to study this problem. However, despite such efforts, we agree with [18] that a common understanding of what is meant by privacy is still missing. As a consequence, there is a proliferation of many completely different approaches of privacy-preserving data mining: some aim at individual privacy, i.e., the protection of sensitive individual data, while others aim at corporate privacy, i.e., the protection of strategic information at organization level. Figure 2 shows a taxonomy tree of privacy-preserving data-mining

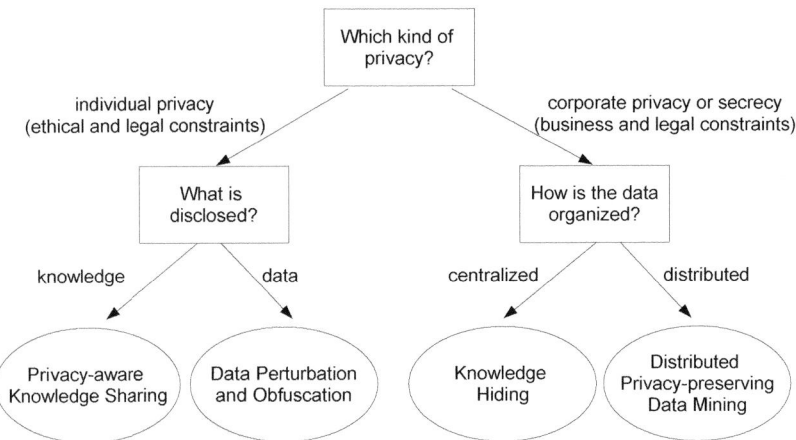

Fig. 2. A Taxonomy of Privacy-Preserving Techniques

approaches in literature. Techniques are classified depending on which kind of privacy needs to be preserved and in which setting data and patterns are.

While security in ubiquitous computing was investigated in the context of sensor network security and security in embedded systems, privacy in ubiquitous computing was studied from the perspective of location privacy. Location privacy is concerned with privacy breaches due to location information, and anonymity of users has been considered the goal to reach. Therefore spatio-temporal anonymity methods need to be addressed. Anonymity can be seen as a preprocessing step for ubiquitous knowledge discovery. We also identify the issue of trust in ubiquitous environments as an important research topic between security and privacy.

The inference problem was addressed in the statistical database community. Now with many different types of heterogeneous data sources, this problem is even more difficult to solve. Access control is necessary but it may not be enough, we also need to keep track of how the data is being used after its collection.

Privacy-Preserving Collaborative Knowledge Discovery. Distribution of the data introduces new problems and makes the privacy more challenging. The scenario is the following: there are data holder sites, and they do not wish to share their data with the others due to regulations or due to the commercial value of the data; another reason is that merging all these data sources may itself be a privacy threat. Therefore individual sites are not willing to disclose their data. So the problem can be phrased as "How to collaboratively obtain the global data mining models without disclosing our data?". The sub-area of research known as distributed privacy-preserving data mining tries to address this problem.

Two frameworks have been investigated for Privacy-Preserving Collaborative Knowledge Discovery:

- one approach is to employ a "trusted third party" (TTP), in which the actual computation is performed at an external party, on which all the participants rely to a great extent.
- the other approach is based on "secure multi-party computation" (SMC), i.e., there is no TTP and its functionality is simulated by participants of the protocol, resulting in a more distributed computation. In addition, other models are needed to formulate the behaviors of the participants, which must be considered as an assumption in the protocol design. A malicious party may involve extra-protocol activities to deceive other parties and to get an upper hand during the protocol. Its goals may range from learning inputs of other parties to preventing other parties from learning the true result of the protocol.

The implementation of the TTP model is, in general, quite trivial, just requiring encryption of all the messages. In the sequel, we will focus on SMC. Secure multi-party computation is a protocol in which participants of the protocol can compute some common function of local inputs, without revealing the local inputs to each other. Each participant participates in the protocol by supplying one or more inputs. At the end of the protocol, only additional information each participant learns is the result of the computation and nothing else. For instance, the participants learn nothing about the inputs of other participants.

Table 1. Description of a secure multi-party computing protocol that calculates a secure difference (see also Fig. 3)

Trace of the SMC Protocol in Fig 3 for computing a-b
step 1: Site A sends k to TP1 and (a-k) to TP2
step 2: Site B sends l to TP1 and (b-l) to TP2
step 3: TP1 computes (k-l)
step 4: TP2 computes (a-k) - (b-l)
step 5: site DM computers (a-b)

In the general case, the task is to securely evaluate a function $y = F(x_1, x_2, ..., x_n)$, with inputs x_i's are coming from the participants. The SMC protocol must be fair in the sense that each participant will have access to the result, y, and no participant can prevent others from retrieving the result. A generic SMC protocol that can be applied to any function usually renders an inefficient solution to the problem. Therefore, tailored solutions for specific applications are preferred. SMC protocols based on secret sharing could be the solution to the scalability problem since they do not require expensive encryption operations. In Figure 3 reader can see the data flow for an SMC protocol for comparing the

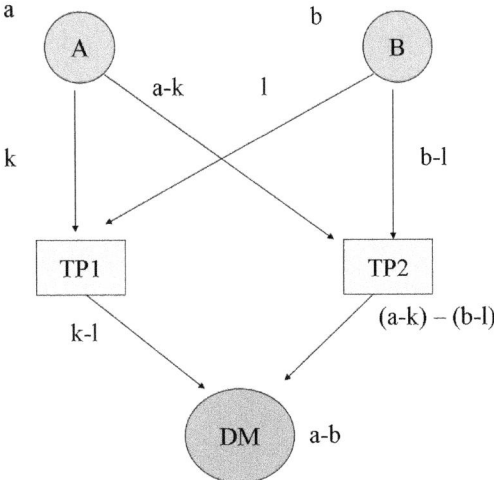

Fig. 3. Secure Difference Calculation with SMC Using Secret Sharing

difference of two numbers a and b where site A has a and site B has b. At the end site DM can compute $a - b$ without knowing either a or b. Secure computation of the difference between two numbers is the basis of calculating the Euclidean distance (a popular distance metric for clustering) of objects in different sites without each site revealing its data.

A data model which definitely fits communications in ubiquitous computing applications is data stream. Unfortunately data streams haven't been studied enough in the privacy-preserving knowledge discovery community. One work [19] on privacy-preserving sequential data, which can be seen as an approximation of the (possibly infinite) data streams, as been recently proposed. Authors show how frequent subsequences in the data can be extracted from a number of data servers using SMC, without joining the data for privacy reasons.

Location Privacy. In the context of ubiquitous computing, another important privacy issue is about location of users, and the risk of tracing them. Location privacy is a quite established research area in privacy-preserving location-based services. Such services are usually a simple function where the input is the (often approximated) location of the user. The aim is on allowing such a service without disclosing the exact position of the user. We are not aware of location-based services of data-mining results.

Definition k-Anonymity: A table T is k-anonymous with respect to a set of attributes QI if each record in the projection of T on QI appears at least k times.

Several papers have been published on privacy-preserving location-based services, exploiting the concept of k-anonymity [20,21]. Most of them rely on a Trusted Third Party (TTP), which is a middleware between the final user and the untrusted service provider. The TTP server typically runs a cloaking algorithm, releasing to the service provider the approximated positions of the users requiring some location-based services. Locations are therefore anonymized by releasing regions in which at least k users are requiring (or may require) a service [22,23]. Each service is usually associated with a pseudonym, and only the TTP server can map pseudonyms with real identities.

A more complex problem is the one shown in the work by [24]. The work analyzes privacy issues when location-based services that requires authentication or the history of required services (called personal history of locations, PHL). Anonymizing independently each location is not enough in this case, since users can be traced by using several service requests. Authors define Historical k-Anonymity, i.e., a sort of anonymization of location sequences.

3   Open Problems

Ubiquitous computing has three important features: high distribution, pervasion into our daily life, and mobility. Battery-life, memory and network bandwidth are considered to be the main resource constraints in ubiquitous computing. The data mining efforts that are being conducted in the context of distributed knowledge discovery should be extended to high degree of distribution, where the number of sites is in the order of thousands. Current distributed mining as well as privacy-preserving distributed mining efforts are not scalable enough to handle high data distribution in ubiquitous systems. Knowledge discovery in the grid would be one of the most relevant research areas in this context. Knowledge discovery in mobile environments should consider the space and time dimensions of the data which is still an ongoing research area (see GeoPKDD project[4]). Spatio-temporal data mining research will therefore be one of the most relevant research areas. Spatio-temporal knowledge discovery is still in its infancy, hence privacy in spatio-temporal knowledge discovery is yet to be addressed.

When considering privacy and security impacts in ubiquitous knowledge discovery, there are two aspects which require attention. The first is how to assess and mitigate security and privacy threats posed by data mining in ubiquitous environments. The second is how to use knowledge discovery to enhance the security and privacy in ubiquitous environments. In what follows, we will consider each of these aspects.

3.1   Privacy and Security Threats Caused by Knowledge Discovery

There is no protection against an almighty adversary. Therefore, an explicit adversary model is needed. This model describes the capabilities of the adversaries who might try to compromise privacy, and against whom a system is required to be secure. Given an adversary model, one can apply the appropriate security measures and infer the protected areas of the system, which are unconditionally secure against the modeled adversary. Protected areas are the foundation of any

---

[4] http://www.geopkdd.eu/

security mechanism. An explicit adversary model is particularly important for ubiquitous systems, because their components are more vulnerable to physical and logical attacks than classical computing devices. The adversary model can be interpreted as a condition on the environment in which the system operates securely. Many adversarial models may be possible for knowledge discovery applications in UbiComp environments. For example, one model may consider attacks that the adversary launches on knowledge discovery applications. Another model may consider an adversary who uses knowledge discovery tools as a weapon to launch attacks on ubiquitous systems.

Avoiding a central data warehouse would be the first step for added security and privacy protection, as large data collections are prone to attacks, and data spills are like a gold mine for malicious attackers. However, does knowledge discovery in ubiquitous systems pose additional privacy and security threats, such as inference of private information? The inference problem was addressed in the statistical database community. In ubiquitous computing there are multiple heterogenous data sources, which makes the inference problem even more difficult to solve. While existing work identifies some privacy and security threats introduced by the process of knowledge discovery or its output, this problem is still largely considered an open question.

Database security is an established area of research, especially in centralized environments where access control techniques are developed for database security. Since we are interested in ubiquitous knowledge discovery, the usage of data will mean the application of data-mining techniques on the data. Therefore, enforcing access control on data-mining results will also be needed. In addition, access control techniques need to be extended with space and time dimensions to meet the security and privacy requirements of ubiquitous systems. Moreover, while access control is necessary at the data level, it may not be sufficient to ensure privacy. Privacy considerations may require to give the users more control over their data and the usage of their data. This can be done by setting data owner preferences and data requester policies. Enforcement of these preferences and policies will require to keep track of how the data is being used after it was accessed.

Another interesting research area in the context of privacy and security is anonymity. When we use our mobile device for a location based service, our location is revealed. When we use the easy pay system of the highway, our location, time and frequency of use of the highway are being recorded. Since space and time are two important features of ubiquitous data, spatio-temporal anonymity methods need to be addressed in this context. In fact, location based services and other location related applications will only succeed if anonymity issues are resolved. Anonymity can be seen as a preprocessing step for ubiquitous knowledge discovery. Security of wireless sensor networks is a relevant area that can provide us with some initial research results.

To summarize, some of the questions that should be addressed in future research are:

- What are the various adversary models that should be considered?
- When do mined patterns preserve privacy?
- Does Ubiquitous Computing pose additional problems to privacy-preserving data mining?
- Is privacy-preserving data mining possible when transactions are insecure?
- How can knowledge discovery tools be applied while ensuring anonymity is maintained?

3.2 Enhancing Privacy and Security through Knowledge Discovery

While knowledge discovery can be used as a weapon by a malicious entity, scheming to compromise security and privacy in ubiquitous systems, it can also be used as a defense aid by the owners and users of the same systems. Some possible goals and questions regarding the use of knowledge discovery to this end are:

- Perceptivity: identifying that the context/environment is corrupted
- How to develop dependable UbiComp environments using knowledge discovery
  - Availability: identify that a property of the system is no longer guaranteed
  - Interactivity: identify that a ubiquitous device is unresponsive to environmental signals
- How can knowledge discovery help to prevent/react against adversary attacks?
- Can DM help developing validation rules according to an explicit model of trust?

To prevent interactive attacks, aimed at denial of service, it is important to develop the appropriate DM methods for the analysis of the stream of requests in order to identify possible misuse of the UbiComp system's augmented facilities. Since reaction time is a critical factor, the efficiency of the DM method should be considered a fundamental requirement when detecting interactive attacks. Real-time data mining can become a new active research area, where efforts on distributed data mining, knowledge discovery from data streams and on-line learning algorithms do synergistically meet.

4 Emerging Challenges and Future Issues

We can now envision a future when embedded devices will be integrated more into our daily lives, they will have more computing power and storage capacity, and they will be able to communicate with each other through wireless technology. Moreover, advanced applications will be built on such devices which will cause another phase shift in the sense that, the UbiComp will be armored with knowledge discovery tools that will allow decision making in UbiComp. When intelligence through data mining in UbiComp is introduced, we will have KDubiq. Before we reach that phase, the privacy and security issues in KDubiq need to be investigated and possible attacks, and privacy breaches should be discovered to make such a technology publicly acceptable. Current technology on privacy protection is not scalable to highly distributed environments and it is prone to privacy breaches in terms of highly heterogeneous data sources.

References

1. Saltzer, J.H., Schroeder, M.D.: The protection of information in computer systems. Proceedings of the IEEE, 63–69 (1975)
2. Perrig, A., Stankovic, J., Wagner, D.: Security in wireless sensor networks. IEEE Computer (June 2004)
3. Diffie, W., Hellman, M.E.: New directions in cryptography. IEEE Transactions on Information Theory 6(22) (1976)
4. Stajano, F.: Security for Ubiquitous Computing. John Wiley and Sons, Chichester (2002)
5. Robinson, P., Zimmer, T.: Structured analysis of security in ubiquitous computing (2007)
6. Covington, M., Long, W., Srinivasan, S., Dey, A., Ahamad, M., Abowd, G.: Securing context-aware applications using environment roles. In: ACM Workshop on Role Based Access Control (2001)
7. for Standardization, I.O.: ISO9 ISO IS 15408 (1999), http://www.iso.org/iso/en/CatalogueDetailPage.CatalogueDetail?CSNUMBER=40613
8. Pfitzmann, A., Hansen, M.: Anonymity, unlinkability, unobservability, pseudonymity and identity management a consolidated proposal for terminology (May 2006), http://dud.inf.tu-dresden.de/Anon_Terminology.shtml
9. Verykios, V.S., Bertino, E., Fovino, I.N., Provenza, L.P., Saygin, Y., Theodoridis, Y.: State-of-the-art in privacy preserving data mining. ACM SIGMOD Record 1(33) (2004)
10. McGraw, G.: A new class of wireless attacks (2001), http://www.linuxsecurity.com/articles/network_security_article-3858.html
11. Wang, H., Huang, J., Qu, Y., Xi, J.: Web services: problems and future directions. Journal of Web Semantics 1 (2004)
12. Orr, R., Abowd, G.: The smart floor: A mechanism for natural user identification and tracking. Technical report, Georgia Institute of Technology (2000)
13. Lee, W., Stolfo, S., Mok, K.: Adaptive intrusion detection: A data mining approach. Artificial Intelligence Review 16 (2000)
14. Wood, A., Stankovic, J.: Denial of services in sensor networks. IEEE Computer (October 2002)
15. Clifton, C., Marks, D.: Security and privacy implications of data mining. In: Proceedings of the 1996 ACM SIGMOD International Conference on Management of Data, SIGMOD 1996 (1996)
16. O'Leary, D.E.: Knowledge discovery as a threat to database security. In: Knowledge Discovery in Databases. AAAI/MIT Press (1991)
17. Agrawal, R., Srikant, R.: Privacy-preserving data mining. In: Proceedings of the 2000 ACM SIGMOD International Conference on Management of Data (SIGMOD 2000), pp. 439–450 (2000)
18. Clifton, C., Kantarcioglu, M., Vaidya, J.: Defining privacy for data mining. In: Natural Science Foundation Workshop on Next Generation Data Mining, pp. 126–133 (2002)

19. Zhan, J.Z., Chang, L.W., Matwin, S.: Privacy-preserving collaborative sequential pattern mining. In: Proceedings of Workshop on Link Analysis, Counter-terrorism and Privacy (2004)
20. Samarati, P., Sweeney, L.: Generalizing data to provide anonymity when disclosing information. Technical report, SRI International (1998)
21. Sweeney, L.: k-anonymity: a model for protecting privacy. International Journal on Uncertainty Fuzziness and Knowledge-based Systems 5(10) (2002)
22. Gedik, B., Liu, L.: Location privacy in mobile systems: A personalized anonymization model. In: Proceedings of the 25th International Conference on Distributed Computing Systems (ICDCS 2005), pp. 620–629 (2005)
23. Mokbel, M.F., Chow, C., Aref, W.G.: The new casper: Query processing for location services without compromising privacy. In: Proceedings of VLDB 2006 (2006)
24. Bettini, C., SeanWang, X., Jajodia, S.: Protecting privacy against location-based personal identification. In: Proceedings of 2nd VLDB Workshop on Secure Data Management, SDM (2005)

A Human-Centric Perspective on Ubiquitous Knowledge Discovery

Bettina Berendt[1] and Ernestina Menasalvas[2,*]

[1] Department of Computer Science, Katholieke Universiteit Leuven
[2] Facultad de Informática, Universidad Politécnica de Madrid
Bettina.Berendt@cs.kuleuven.be,emenasalvas@fi.upm.es

1 Introduction: What Are the Right Questions to Ask for a Human-Centric View of KDubiq?

The relevance of a human-centric view for KDubiq in particular or knowledge discovery (KD) in general arises from the comprehensive nature of KD. KD is more than the application of algorithms – it encompasses the whole process of turning data into knowledge: business/application understanding, data understanding, data preparation, modelling, evaluation, and deployment [1]. People play a pivotal role in this process: people create data, data and knowledge are about people, and people are (or some of them should be) the ultimate beneficiaries of the discovered knowledge. As shown in the previous chapters, ubiquity adds more dimensions to KD; this chapter analyzes how these dimensions affect the user perspective of KDubiq systems.

In order to take the people-centric view we aim for, we consider results from two areas of Computer Science and Cognitive Science that strive towards a human-centric view of information systems. *Human-computer interaction (HCI)* is the study of interaction between people and computers, which occurs at the user interface (hardware and software). *Cognitive modelling (CM)* is the use of computational methods to model mental representations and processes and the behaviour resulting from them. Many HCI decisions depend on the cognitive processes that the user goes through or is meant to go through during interaction; in this sense, cognitive modelling is an essential part of HCI.

A full picture of HCI and CM for KDubiq may be obtained by considering the impact of each feature of KDubiq for each stage of KD.

The features of KDubiq to be investigated are (as defined in Chapter 1, Section 2): the presence of a population of devices with computing capabilities (in the following: *device ubiquity*), an extreme distribution of data (*data ubiquity*), a continuous flow of incoming data (*data flow*), and a rich semantics of data due to the heterogeneity of the involved agents. As in Chapter 4, it is helpful for our purposes to distinguish between sources of heterogeneity; we will refer to

---

* With contributions from the WG6 members:

www.cs.kuleuven.be/~berendt/HCI-ubiq

M. May and L. Saitta (Eds.): Ubiquitous Knowledge Discovery, LNAI 6202, pp. 90–107, 2010.

computing heterogeneity (e.g., of sensors) and *people heterogeneity*. In line with the present chapter's people-centric view, we will go beyond Chapter 4 in our treatment of people heterogeneity: whereas that chapter states *that* people are heterogeneous, we ask *in what ways* they are heterogeneous and *which specific effects* this has on KDubiq. *Privacy* will also be investigated here, in a people-centric view that complements the data-centric one of Chapter 5.

The KD stages to be investigated are in principle all the standard stages of KD (in the sense of CRISP-DM [1] as used above). However, the CRISP-DM stages structure the process of KD by the activities of "analyst users", whereas most of the people interacting with a KD or KDubiq system will be "end users" for whom the application is made (patients in Chapter 1, Section 6.1, music lovers in Chapter 1, Section 6.2, car drivers and pedestrians in Chapter 5, Section 2, etc.). They generally perceive only a coarsened version of the CRISP-DM stages, which can be described as *data collection, data processing, information presentation and interaction* in all stages, and *evaluation*. Prior to these, and in cases of good software engineering, end users are also the (knowing or unknowing) subjects of an earlier phase: that of requirements analysis. We integrate this stage to ensure that a comprehensive view of users, other stakeholders, and their respective interests is taken.

Since KDubiq applications are a type of ubiquitous-computing applications, many observations made in the present chapter also hold true for ubiquitous computing in general. Nonetheless, we aim to emphasize HCI/CM challenges specific to KDubiq (as opposed to UbiComp in general) by our foci on and within these stages: data collection (KD applications need more data than many non-KD-computing applications), data processing (for example, the inferencing inherent to KD generates more privacy problems than the "mere" data storage and retrieval of most non-KD-computing applications) and the impact this has on information presentation / interaction (for example, predictability and transparency become specific usability issues when one deals with an "intelligent", adaptive system).

A people-centric perspective is needed when constructing a KDubiq system because the KDubiq features (as defined in Chapter 1, Section 2) affect all phases of knowledge discovery. The present chapter is a short overview of key challenges, structured by a matrix organisation: What is the impact of KDubiq features on the KD stages? The chapter concentrates on the most relevant combinations of KDubiq features and KD stages (see Table 1).

The chapter thus aims at a systematic, "meta-level" analysis of the HCI/CM challenges faced by the emerging field of KDubiq; rather than at giving an overview of current – and conceivably soon outdated – concrete usage interfaces. We will illustrate the identified challenges, throughout the chapter, by references to the application example ACTIVITY RECOGNITION (see Chapter 1, Section 6.1).

An outlook and visions for the future conclude the chapter.

Table 1. The sections of the current chapter describe the impact of KDubiq features on KD stages. Numbers in the table refer to sections in the current chapter.

KDubiq feature / KD stage	device u. data u.	data flow	rich data semantics		privacy
			computing heterogeneity	people heterogeneity	
Requirements analysis				2.1–2.3	2.4
Data collection	3.1/.2/.4	3.3	3.1	3.3	(and
Inform. present. and analysis	4.2; 3.1		4.1	4.3	through-
Data processing	5.1				out)
Evaluation				6.3	

2   Requirements Analysis

Human-centric algorithm and system development means that the interests of a certain group of people are respected. Requirements analysis of a KDubiq system from a human-centric point raises a number of questions as to who, what, and how.

- *Who* includes: Which people are relevant? Is the total (respecting the interests of a set of people) the sum of its parts (respecting the interests of each one of them), or is it more? How can the interests of a *set* of people be respected?
- *What* is the question of what "interest" means.
- *How* is the question of method: How can interests be elicited from people?

We hold that the answers to the "what" and "how" questions will not differ substantially from the answers in other fields of system design. Therefore, we refer the reader to [2] for details, and concentrate on the "who" in this section.

Even if the "who" question of which people are relevant has been answered, conflicts between different people's interests can arise. This is one of three forms in which conflicts can arise in KDubiq: (i) as contradictory statements about the desired state of the (system's) world, obtained from different users or stakeholders, (ii) as contradictory statements about the actual state of the (system's) world, obtained from different data/knowledge sources, and (iii) as contradictory plans for future action, obtained from the possibility that different things can be concluded or done next. Cases (ii) and (iii) concern aspects of data collection and processing and are therefore discussed in Section 3.2 below. All types of conflicts can arise in any system, but they become more prevalent with the growing number of (human and machine) agents and the growing amount of data in KDubiq. We therefore first turn to the first type of conflicts, that between contradictory statements about the desired system, obtained from different users or stakeholders (Section 2.1), who can be individuals or collectives (Section 2.2), and for which conflicts have to be detected and resolved (Section 2.3).

2.1   Users and Stakeholders

Stakeholders of a system are all persons who have some functional or non-functional (in particular, security or privacy) interest in the system. This encompasses all persons involved in the conception, production, use and maintenance of the system. Stakeholders are not limited to *users*: those who will be using the functionality of the system. This distinction is necessary in ubiquitous systems for two reasons:

1. the concept of users ("actors" in use cases) implies explicit interaction with the interfaces of the given system. Yet, implicit forms of interaction or even no interaction cases are also central to ubiquitous systems.
2. a system may collect, process or even disseminate data which are not related to its users. For example, in customer relationship management systems, (sometimes very sensitive) data about the customers are collected, but the customers are themselves not the users of these systems. Identifying the customers as stakeholders of such systems helps to make customer interests an important part of the system.

   In the ACTIVITY RECOGNITION setting, sensors collect and disseminate information about the user and her environment for the service being provided, but this information may later be used for other purposes not under the user's control.

In human-centric ubiquitous systems it is therefore indispensable to consider the functional, knowledge and privacy/security interests of all the stakeholders. To do this, we borrow from two existing approaches.

Viewpoint-oriented requirements analysis recognizes that one needs to collect and organize requirements from a number of different viewpoints – those of the different stakeholders [3]. Here, we concentrate on privacy-related viewpoints as a key challenge for ubiquitous computing and KDubiq. A viewpoint approach to security has been suggested in *multilateral security* engineering, see [4,5] for references and applications to human-centric KDubiq.

2.2   Individuals and Collectives

In addition to broadening the view from users to stakeholders, in KDubiq one must also go beyond individual users (the traditional focus of computer science). *User communities* are groups of users who exhibit common behavior in their interaction with an information system [6,7]. Basic assumptions of data processing may have to be adapted when addressing groups rather than individuals [8].

Building systems for groups also needs to take into account not only the best achievable technological possibilities, but also the economic realities of users. This may result in lowest-common-denominator solutions (in the case study on ACTIVITY RECOGNITION, if systems to help impaired people are to be built, it is especially important to end up with a solution feasible for the majority of them).

2.3 Conflicts and Conflict Resolution

Conflicts (interferences between different party's activities and interests) are recognized in the trade-offs between functional requirements and non-functional requirements (safety, security, usability, performance, cost, interoperability etc.). Ubiquitous computing magnifies both the capacities of systems and the number of stakeholders and thereby the potential for conflicts in its environments. It is now accepted in requirements engineering that conflicts should not be avoided, and that instead methods for dealing with them have to be developed [2].

Many methods for negotiating or resolving conflicts during an iterative requirements engineering process exist [9,10,11,12], but conflicts may arise during the later design stages, or even after deployment. Thus, conflict management needs to be a part of not only system development but also maintenance.

2.4 Privacy

The importance of privacy in KDubiq is described in Chapter 5; in system development, it is usually considered a non-functional requirement. But what is privacy, in particular from a human-centric viewpoint? In many laws as well as in the data mining community, the concept of privacy is generally translated into (personal-)data protection.

Personal data is "any information relating to an identified or identifiable natural person [...]; an identifiable person is one who can be identified, directly or indirectly, in particular by reference to an identification number or to one or more factors specific to his physical, physiological, mental, economic, cultural or social identity" [13], Art. 2 (a). *Data privacy* and *informational self-determination* denote restrictions on the collection and processing such data, and control of access to them.

These definitions focus attention on data (as opposed to people), and on the detection of – or inference towards – "the" identity as the problem. However, the privacy literature suggests that a people view of privacy involves not one but many identities, that there are concerns over profiles independently of identification, and that context is all-important. Phillips [14] distinguishes four kinds of privacy: (1) *freedom from intrusion*, also known as the "right to be let alone"; (2) *construction of the public/private divide*, the social negotiation of what remains private (i.e. out of the public discourse) and what becomes public; (3) *separation of identities*, which allows individuals to selectively employ revelation and concealment to facilitate their social performances and relationships; and (4) *protection from surveillance*, where surveillance refers to the creation and managing of social knowledge about population groups. These four kinds of privacy are neither exhaustive nor mutually exclusive, but they help in understanding which aspects may be protected or threatened by emerging technologies (see [14] for the relations between different privacy-enhancing technologies like anonymizers or identity management systems and the four kinds of privacy).

In addition to these different foci, privacy can be regarded as a basic and inalienable human right, or as a personal right or possession over which the individual has free discretion. When regarded as the former, privacy has to be regulated through laws, political decisions, and ethics. When regarded as the latter, it becomes a matter of utility, taste and freedom of choice, see [4,5] for extended discussions.

3   Data Collection

Human-centric issues of data collection in knowledge discovery usually concern "personal data". These include relatively stable properties of the users such as age, profession, or place of residence, and transient properties such as task, mood, or time pressure.

Personal data may be valuable or essential for system adaptation (and thus usefulness), but it may be cumbersome for the user to input these data. Therefore, it is often considered desirable to let the system record such data without the user noticing it. However, users may not want to share such data, and object to non-noticeable ways of recording them. In addition, non-user stakeholders' data may be implicated.

Therefore, privacy becomes a key concern. Four important aspects have to be analyzed:

- the impact of data and device ubiquity,
- conflicts resolution,
- people heterogeneity,
- context and context measurement.

3.1   Impact of Data and Device Ubiquity and of Computing Heterogeneity

These features of KDubiq pose new challenges, in particular for privacy. We will concentrate on location privacy with respect to ubiquitous data. We also use location privacy to demonstrate the increasing difficulties of knowing about and controlling data about oneself in ubiquitous environments [4].

Location is most easily defined as the triple (ID, position, time). Location data are particularly sensitive information because of the specific characteristics of space. As one cannot not be in a location at any time (one cannot "opt-out of being somewhere"), the impression of lacking self-control and comprehensiveness of surveillance is particularly pronounced. In addition, spatial and spatio-temporal data allow many inferences because of rich social background knowledge.

The preservation of location privacy in ubiquitous environments is extremely difficult (see Chapter 5 for proposals). Trajectories, i.e., temporally ordered series of location-triple observations with the same ID, coupled with a-priori knowledge on places and social contexts, can be used to infer individual identities (violating freedom from intrusion), to link profiles of individuals (violating separation of identities), or to classify persons into previously defined groups (surveillance). In the ACTIVITY RECOGNITION example, any of these may result from, for example, the simple identification of doctors (and their specialties) that the user consults frequently, or friends that the user visits regularly.

3.2 Conflicts and Conflict Resolution

The multitude of devices in ubiquitous environments that may independently of each other gather data and draw inferences on the same entities may also lead to conflicts of type (ii) in the sense of Section 2: contradictory statements about the actual state of the (system's) world. This form of conflict has been extensively studied in AI belief revision systems, often borrowing not only from argumentation logic, but also from CM. Conflict resolution strategies generally rely on heuristics about relative trustworthiness. For example, heuristics can help decide which inferences towards either of the conflicting beliefs are more trustworthy; sample heuristics state that authoritative sources are believed more strongly, and more specific information is considered more accurate than more general information (see [15] for a comprehensive analysis and solution proposals in the context of ubiquitous user modelling).

3.3 Impact of People Heterogeneity

People heterogeneity has two major impacts on data collection (for references and more detail, see [16]):

1. language, culture, and other aspects of people heterogeneity may influence whether people have access to data and knowledge in the first place [16].
2. these factors may also influence whether and to whom people want to disclose/share personal data, and which data they regard as "harmless" or "sensitive" in this respect.

Whereas language predominantly affects the capacities of sharing data and knowledge, culture has a major impact on users' willingness to share information. In many studies, "culture" is operationalized in terms of Hofstede's [17] "cultural dimensions" that rank countries for example by the closeness or looseness of ties within society (*collectivism – individualism*) or the degree of acceptance of unequal power distributions within society (*power distance* – high if unequal power distributions, i.e. strong social hierarchies, are accepted; low if 'flat hierarchies' are the accepted norm).

For example, collectivistic cultures – in contrast to individualistic ones – tend to strongly differentiate between ingroup and outgroups. Various studies indicate that, as a consequence, collectivistic cultures fear fraud and moral damage through the information that is provided on the Internet. Individualistic cultures value private space more than collectivistic cultures and therefore are less willing to provide sensitive information. Members of high power-distance cultures are more willing to provide data than members of low power-distant cultures.

3.4   Context and Data Collection

The multitude of sensors and knowledge sources in ubiquitous systems, and the power of inference in KD systems, together with the perceived importance of HCI and Cognitive Modelling context has led to a large increase in the interest in defining, measuring, and processing context.

Definitions of context in ubiquitous systems focus on what context is about, or on the way in which it is measured. Definitions that focus on what context is about [18] often differentiate between environment (location, time, weather, other properties of the physical environment or computational infrastructure), and persistent or transient properties of the user such as the social environment, preferences, or task, e.g., [19,20,21].

Measuring context is challenging as it is not only related to the factors and the relevant facets in the application context, but also to their representation in the computer.

With regard to user-related factors, most of the time, either context has to be inferred from observations that the system collects over an extended period of time, or the user has to be directly asked. Privacy and security of the user also restrict the measurability of some situational factors [22] that cannot therefore be obtained even though there may be sensors that can reliably measure it.

As an example, proposals for adding more sensors for purposes such as AC-TIVITY RECOGNITION aim at better context measurement, but this may be ambiguous. For instance, does the measurement that the user is not moving mean that she had an accident and needs help, or simply that she is taking a nap that deviates from the normal daily routine?

4   Information Presentation and Interaction

The field of human-computer interaction has – by definition – focussed strongly on issues of interaction, with usability being the key goal. The presentation of information is one of the core elements of interaction. Therefore, in this section we investigate basic usability challenges in general and usability challenges that become particularly prevalent in user-adaptive systems (including systems that deploy KD towards user adaptation) and in KDubiq systems.

4.1   Basic Usability Challenges

Usability is the effectiveness, efficiency, and satisfaction with which specified users can achieve specified goals in a particular environment (ISO 9241). Usability goals [23] are desirable properties of interactive systems that correspond to general usability principles. The goals are: *predictability*, the extent to which a user can predict the effects of her actions; *transparency*, the extent to which a user can understand system actions and/or has a clear picture of how the system works; *controllability*, the extent to which the user can prevent particular actions or states of the system if she has the goal of doing so, and *unobtrusiveness*, the extent to which a system does things "in the background" not reducing the user's ability to concentrate on her tasks. The importance of these four principles can be observed for the ACTIVITY RECOGNITION case study.

These goals are important for any interactive system, but they may be more difficult to achieve in user-adaptive systems: predictability and transparency may be lower because the system does not always respond in the same way (since it adapts to the user), controllability may be an additional issue since a user-adaptive system by design takes control of many adaptation processes (rather than for example a customizable system where the user controls the adapation), and unobtrusiveness is faced with new challenges because much more knowledge gathering about the user is necessary in order to adapt than in non-adaptive systems.

We will address ubiquity-specific issues of these four usability goals. (The goals are complemented in [23] by two others: Privacy, which we treat throughout the chapter, and Breadth of Experience, which we cannot address here for reasons of space.)

4.2   Impact of Device and Data Ubiquity and of Computing Heterogeneity

In the following, we first address the implications of the newly-found relevance and measurability of context (see Section 3.4) on usability as well as the implications of the new emphasis on system initiative. We will also analyze whether ubiquitous systems change the very nature of "interaction".

Context and awareness. Awareness in its most general sense is defined as "having or showing realization, perception, or knowledge" (Webster dictionary definition). The notion of awareness most often found in ubiquitous computing (*context awareness*) extends these ideas to systems being aware of others: "Computers can both sense, and react based on their environment. Devices may have information about the circumstances under which they are able to operate and accordingly react based either on rules or because an intelligent stimulus. Context-aware devices may also try to make assumptions about the user's current situation." [24].

From pull to push: how much information and when? Context awareness is about systems becoming more perceptive. But there are still some issues that remain unexplored such as how to adapt to changes in context while at the same time adhering to the principle of predictability (and thus transparency).

The challenge in ubiquitous environments of designing systems that react to context is due to the fact that not only location of the user changes but also other contextual variables. One central theme is whether the information should be provided to the user or being asked by the user. Information provision can vary between *pull*, the information flow being triggered (and therefore expected) by the user and *push*, the occurrence of information flow even when it is not expected by the user [25].

Using a spatial model can be beneficial to represent information flows, and the information flow's potential for task interruption and annoyance can be reduced. This has been already analyzed in [26] showing that it is useful to consider a design space that maps out, for a given presentation device, to whom the information is directed (public vs. private) and who the information reaches (public vs. private). By means of this model it is also possible to define areas of influence for "presentation" devices.

Control. Context-aware computing aims to facilitate a smooth interaction between humans and technology. However, few studies show how users perceive context-aware interaction. This is noted by Barkhuus and Dey in [27], which contains an analysis to compare users' responses towards applications, showing that participants feel a lack of control as the autonomy of interactivity is increased. The authors conclude that users accept larger degrees of autonomy from applications when the application's usefulness is greater than the cost of limited control.

Another interesting aspect related to control is the user controlling the information the system has of her, which again is a challenge in KDubiq as users must be in control of not only the response of the system but also of the information that has been collected and preprocessed. In the ACTIVITY RECOGNITION example, sensors of the system are recording every bit of information related to the user, her context and her actions. Thus the user not only should know exactly the information that is being gathered but she should also know the degree to which this information can be used for her or for third parties to get a better service or added value. Do not forget that information is not staying at her device but is shared with other devices. It is at this moment that privacy is in danger as the information is not under the user's control any longer.

Implicit interaction. The wording "usability challenges" that we have used throughout this section, suggests that there is something that the user perceives she is using (and then assesses this act of using for its qualities). However, with the increasing emphasis, in ubiquitous computing, on "unobtrusive" data gathering, often by a multitude of "invisible" sensors, and distributed processing of these data whose complexity is generally hidden from the user, the very notion of interaction changes. The provision of input and the appreciation of output were (largely) intentional and conscious user acts in traditional computing environments. Nevertheless, they are dwarfed or even supplanted by system acts that may neither reflect a user's intention nor even be noticed by her. This *implicit interaction* [28] not only changes the notion of interaction, but may even change the notion of user and/or imply that many more non-user stakeholders are affected by system actions. Viewed in a usability context, this implies, even more strongly than context awareness per se, that systems may become more unobtrusive (or "calm"), but that transparency and predictability are significantly reduced (e.g., [29, p. 70]).

Note that unobtrusiveness is *not* a necessary consequence, however: As the observations in Section 4.2 have shown, proactive context-aware systems may actually be *more* obtrusive if designed badly. This is a mandatory principle for the case of ACTIVITY RECOGNITION devices, where the system has to help the user, but the user has to have the control of stopping the system gathering and learning from her behaviour when wished. Viewed beyond a usability context, implicit interaction draws attention to the necessity of rules that go beyond contractual relations between a service provider and a user, and that observe the external effects on other stakeholders.

4.3   Impact of People Heterogeneity

People heterogeneity also influences users' preferences for information presentation and their evaluation of information (see [16] for more details).

Language predominantly determines how easy it is to access data and knowledge. Ease of use is a determinant of usefulness, attitude towards an information system, and satisfaction. It can therefore be assumed that information in a user's native language leads to a better evaluation.

Culture affects people's preferences for certain forms of data presentation. For example, users from high power-distant countries appear to prefer a hierarchical form of knowledge presentation more than members of low power-distant countries. Preferences for classification schemes are also affected by cultural differences.

5   Processing

Device ubiquity implies that processing may need to operate on distributed data, implying also that inference itself may be distributed over different nodes. Maintaining privacy becomes an even stronger concern in this stage of KD. This is the topic of a group of "privacy-preserving data mining" (PPDM) approaches, which are dealt with in Chapter 5. We focus on two issues:

- how to create conditions for sharing and joining the heterogeneous data (1): interpreting data models;
- how to create conditions for sharing and joining the heterogeneous data (2): communicating data models;
- how to deal with conflicts that may arise during processing.

5.1   Interpreting and Processing Distributed Data: The Semantic Approach

In the area of distributed information, different users might have quite different views of a data collection. Data heterogeneity can produce models that carry only minimal semantics and are often not fully useful for the user.

Standards like XML and RDF decrease syntactical heterogeneity, but semantic heterogeneity is still a challenge. Ontologies are an important building block for overcoming this challenge. Folksonomies (see Chapter 4), on the other hand, do not depend on common global concepts and terminology and replace ontologies in Web 2.0 approaches, allowing users to assign arbitrary tags to content.

The Situation Semantics [30] offers a framework for conceptualizing everyday situations in which the actual world can be thought of as consisting of situations and in turn situations consist of objects having properties and standing in relationships. Nevertheless, semantically interpreting and processing data remains a challenge in ubiquitous environments.

5.2   Communicating about Distributed Data and Models

The sharing of data and knowledge among processes in an inherently distributed framework is another challenge that has to be tackled.

The user markup language UserML [31] and its corresponding ontology UserOL were proposed to contribute a platform for the communication about partial user models in a ubiquitous computing environment.

GUMO [32] makes the uniform interpretation of distributed user models possible due to its capability to collect users' attributes, context conditions and information about the enviroment and the devices.

Consequently GUMO together with UserML represents a promising area for the semantic integration in ubiquitous user-adaptive applications.

5.3   Conflicts and Conflict Resolution

The multitude of devices, processing units and stakeholders in ubiquitous environments may lead to conflicts of type (iii) in the sense of Section 2: contradictory plans for actions. In AI, this form of conflict has been studied in multi-agent systems, more specifically in systems with non-cooperative settings. The standard assumption is that such agents behave according to the principles of rational decision making, such that conflict resolution is approached by mechanism design in the sense of economics. This is the game-theoretic approach in which mechanisms and negotiation protocols are designed so as to ensure the mutually optimal outcome. Thus, communication between agents is generally emphasized as the central organising principle [33].

However, such a proposal is not really practical in a ubiquitous computing context. The demands for continuity of service across a broad range of technological environments require strategies that minimize real-time communication between devices for decision making, and at the same time allow agents to use broad-band communication for updates and reconfiguration at 'idle' times.

Users (or person-related devices like PDAs) as well as environments can be described as agents that can prioritize, plan, and learn. Prioritizing may be a part of design (such as emergency information being given precedence over entertainment), or it may be a result of real-time negotiation in the event of a conflict. First, such systems respect the constraints of ubiquitous processing and real-time action; second, they can learn from the interactions they go through. In this type of architecture, conflict resolution is partly the result of hierarchical planning (without communication) based on modular behaviour-based AI [34], and partly the result of negotiation (with communication) as in multi-agent systems. This type of architecture is well-suited to KDubiq for two reasons. First, such systems respect the constraints of ubiquitous processing and real-time action; second, they can learn from the interactions they go through.

6   Evaluation

In software development, evaluations are used to determine the quality and feasibility of preliminary products as well as of the final system. User satisfaction is one of the aspects to be evaluated, and it has been shown how the question of "who" the user is affects KDubiq systems. Consequently, the impact of KDubiq features has to be taken into account when

– evaluating patterns in the sense of KD/machine learning,
– evaluating the deployed systems.

Particular challenges for evaluation arise from

– people heterogeneity.

6.1   Evaluation of Patterns (Before Deployment)

According to CRISP-DM [1, p. 14], at the Evaluation stage "the [obtained] models that appear to have high quality from a data analysis perspective" are analyzed with respect to the application/business objectives. The user plays an important role in this evaluation task. Data and knowledge presentation must respect the limits imposed by the combination of ubiquitous devices and general human perceptual and cognitive limitations.

6.2   Evaluation of (Deployed) Information Systems

Adaptive systems adapt their behavior to the user and/or the user's context. The construction of a user model usually requires making many assumptions about users' skills, knowledge, needs or preferences, as well as about their behavior and interaction with the system. Empirical evaluation offers a way of testing these assumptions in the real world or under more controlled conditions. When constructing a new adaptive system, the whole development cycle should be covered by various evaluation studies, from the gathering of requirements to the testing of the system under development. An evaluation procedure able to uncover failures and maladaptations in the user model is presented in [35]. The model consists of four layers: evaluation of input data, evaluation of inference, evaluation of adaptation decision and evaluation of total interaction.

Nevertheless, as shown in Section 4.2, specific challenges arise for evaluation when interaction becomes implicit. How can an interaction be evaluated by a user when – by definition – it is not noticeable? When there is still a task that the system helps the user perform, the traditional criteria of usability (effectiveness, efficiency, satisfaction) may be adapted for evaluation even when the task is performed with minimal or no perceived interaction in the traditional sense. However, new questions arise when the notion of task becomes fuzzier, and when further stakeholders enter the evaluation.

6.3   Impact of People Heterogeneity

People from different backgrounds evaluate information in different ways (see [16] for references and more details).

Differences in information evaluation are strongly related to differences / compliances in communication styles. In cultures with a clear preference for face-to-face (interpersonal) communication, the reliability of information is attributed to the reliability of its carrier. Technologies are not seen as an equivalent of interpersonal communication and are therefore not as trustworthy, which leads to a higher acceptance of communication technology (e-mail, cell phones, ...) than for impersonal devices (Web sites, eCommerce, ...).

Members of different cultural groups have different approaches towards contradicting information and its evaluation. Particularly the level of power distance appears to have an important impact: members of high power-distant countries tend to more easily accept information unquestioningly.

Users also differ in the way they express their evaluation. For example, it is argued that members of collectivistic cultures often adopt neutral positions (which renders Likert scales with 5 or 7 answer options rather useless).

These differences are likely to have significant effects on their evaluation of a deployed information system. However, to the best of our knowledge, this conjecture has not been tested in systematic ways yet.

7   Emerging Challenges and Future Issues

In the preceding text, we have pointed to many open problems and challenges. Further open problems comprise the cells in Table 1 that we could not describe here for reasons of space. We conclude with a summarizing sketch of key research issues, again structured by the KD stages. The realization of each of them will contribute to the vision of truly human-centric KDubiq systems for the future.

Requirements. Ubiquity enlarges the definition of target groups of a system to include users and other stakeholders as well as individuals and communities. Consequently, methods to elicit possibly conflicting interests of the target population are required. Conflicts arise mainly due to data coming from different users or stakeholders, obtained from different data/knowledge sources. As conflicts should not be avoided, methods for dealing with them have to be developed. Existing requirements engineering methods that investigate different viewpoints [3] and requirements interactions [12] as well as methods for negotiating or resolving conflicts [9,10,11] are useful starting points for such endeavours.

Data collection. Data collection from ubiquitous users must cope with two major problems: challenges of obtaining data and challenges of their representativeness.

Data collection efforts that rely on users' self-reports need to consider that users differ in their ability to provide information as well as in their willingness to share it. Differences in privacy issues and willingness to share information ask for a detailed examination of the extent to which data gathering would constitute an intrusion into the private space. Furthermore, a user's background affects the way opinions are expressed. This needs to be taken into account through either a culturally adapted conception of data gathering tools or through appropriate data processing that considers these differences.

New challenges arise with the new types of data collected from users: while in the past and in many traditional UbiComp systems, these are mostly "personal data", current environments also contain much user-generated content that cannot readily be subsumed under these categories and/or that is decidedly intended for publication and wide dissemination (cf. Chapter 4).

Information presentation and interaction. Research on data presentation forms involves the development of technologies that are able to bridge the gap between different cultures and languages, such as multilingual information retrieval tools. User-centred knowledge discovery should therefore also aim to discover the thresholds where adaptations to the user's linguistic and cultural needs are necessary and where other solutions are more efficient and/or appropriate.

Processing. People heterogeneity leads to heterogeneous data sets due to different contexts. Human-centric knowledge discovery hence requires data processing that takes background knowledge about the users and their context into account. Ontologies and folksonomies have been analyzed as a possible solution that needs to be extended to tackle the above-mentioned context problem. Given the increasing amount of multilingual data, knowledge discovery should also take into consideration research results regarding multilingual information retrieval tools.

Furthermore, the inherent sharing of data and knowledge between entities in this kind of system demands platforms and languages for the communication about partial users models.

Evaluation. Real users are a strong source of information for the knowledge base of the system, and their real behavior offers insight for the intelligent behavior of the system [35]. KDubiq systems demand evaluation methods where the user plays an important role.

In this chapter, we have argued that a human-centric perspective of KDubiq is not only a matter of usage interfaces, but a perspective that spans the whole design process of the system, from the conception of a system through to its deployment. Consequently, the impact of ubiquity – more specifically, of all its features – on the different system development stages was investigated. We conclude that three of the challenges to be addressed in future work are:

- Methods must be developed and/or improved for determining who the user of the system is, and who further relevant stakeholders are.
- Data processing must take background knowledge about the users and their contexts into account in order to respect the heterogeneity of people.
- To better protect the users and stakeholders, methods to examine the extent to which data gathering would constitute intrusions into privacy are needed.

References

1. Chapman, P., Clinton, J., Kerber, R., Khabaza, T., Reinartz, T., Shearer, C., Wirth, R.: CRISP 1.0 process and user guide (2000), `http://www.crisp-dm.org/download.htm`
2. van Lamsweerde, A.: Requirements engineering in the year 00: a research perspective. In: International Conference on Software Engineering, pp. 5–19 (2000)
3. Sommerville, I., Sawyer, P.: Viewpoints: Principles, problems and a practical approach to requirements engineering. Annals of Software Engineering 3, 101–130 (1997)
4. Gürses, S., Berendt, B., Santen, T.: Multilateral security requirements analysis for preserving privacy in ubiquitous environments. In: Proceedings of the Workshop on Ubiquitous Knowledge Discovery for Users at ECML/PKDD 2006, Berlin, pp. 51–64 (2006), `http://vasarely.wiwi.hu-berlin.de/UKDU06/Proceedings/UKDU06-proceedings%.pdf`
5. Preibusch, S., Hoser, B., Gürses, S., Berendt, B.: Ubiquitous social networks – opportunities and challenges for privacy-aware user modelling. In: Proceedings of the Workshop on Data Mining for User Modelling at UM 2007 (2007), `http://vasarely.wiwi.hu-berlin.de/DM.UM07/Proceedings/DM.UM07-proceedin%gs.pdf`
6. Orwant, J.: Heterogenous learning in the doppelgänger user modeling system. User Modeling and User-Adapted Interaction 4(2), 107–130 (1995)
7. Pierrakos, D., Paliouras, G., Papatheodorou, C., Karkaletsis, V., Dikaiakos, M.D.: Web community directories: A new approach to web personalization. In: EWMF, pp. 113–129 (2003)
8. Jameson, A.: More than the sum of its members: Challenges for group recommender systems. In: Proceedings of the International Working Conference on Advanced Visual Interfaces, Gallipoli, Italy, pp. 48–54 (2004), `http://dfki.de/~jameson/abs/Jameson04AVI.html`
9. Easterbrook, S.: Resolving requirements conflicts with computer-supported negotiation. Requirements Engineering: Social and Technical Issues, 41–65 (1994)
10. Boehm, B., Bose, P., Horowitz, E., Lee, M.J.: Software requirements negotiation and renegotiation aids. In: ICSE 1995: Proceedings of the 17th International Conference on Software Engineering, pp. 243–253 (1995)
11. Zave, P., Jackson, M.: Conjunction as composition. ACM Trans. Softw. Eng. Methodol. 2(4), 379–411 (1993)
12. Robinson, W.N.: Integrating multiple specifications using domain goals. In: IWSSD 1989: Proceedings of the 5th International Workshop on Software Specification and Design, pp. 219–226 (1989)

13. EU: Directive 95/46/EC of the European Parliament and of the Council of 24 October 1995 on the Protection of Individuals with Regard to the Processing of Personal Data and on the Free Movement of such Data. Official Journal of the European Communities (L. 281) (1995), http://europa.eu.int/eur-lex/en/consleg/main/1995/en_1995L0046_index.html

14. Phillips, D.: Privacy policy and PETs: The influence of policy regimes on the development and social implications of privacy enhancing technologies. New Media Society 6(6), 691–706 (2004)

15. Heckmann, D.: Situation modeling and smart context retrieval with semantic web technology and conflict resolution. In: Roth-Berghofer, T.R., Schulz, S., Leake, D.B. (eds.) MRC 2005. LNCS (LNAI), vol. 3946, pp. 34–47. Springer, Heidelberg (2006)

16. Berendt, B., Kralisch, A.: From world-wide-web mining to worldwide webmining: Understanding people's diversity for effective knowledge discovery. In: Berendt, B., Hotho, A., Mladenič, D., Semeraro, G. (eds.) WebMine 2007. LNCS (LNAI), vol. 4737, pp. 102–121. Springer, Heidelberg (2007)

17. Hofstede, G.: Cultures and Organizations: Software of the Mind. McGraw-Hill, New York (1991)

18. Dey, A.K.: Understanding and using context. Personal and Ubiquitous Computing 5(1), 4–7 (2001)

19. Wahlster, W., Kobsa, A.: User models in dialog systems. In: Kobsa, A., Wahlster, W. (eds.) User Models in Dialog Systems, pp. 4–34. Springer, Heidelberg (1989)

20. Lieberman, H., Selker, T.: Out of context: Computer systems that adapt to, and learn from, context. IBM Systems Journal 39(3&4), 617 (2000)

21. Schmidt, A.: Potentials and challenges of context-awareness for learning solutions. In: Proc. LWA, pp. 63–68 (2005)

22. Kray, C.: Situated interaction on spatial topics. Dissertations in Artificial Intelligence-Infix, vol. 274 (November 2003)

23. Jameson, A.: Adaptive interfaces and agents. In: Jacko, J.A., Sears, A. (eds.) Human-Computer Interaction Handbook, pp. 305–330. Erlbaum, Mahwah (2003), http://dfki.de/~jameson/abs/Jameson03Handbook.html

24. Wikipedia: Context awareness (2006), http://en.wikipedia.org/wiki/Context_awareness (access date November 21, 2006)

25. Cheverst, K., Smith, G.: Exploring the notion of information push and pull with respect to the user intention and disruption. In: International Workshop on Distributed and Disappearing User Interfaces in Ubiquitous Computing, pp. 67–72 (2001)

26. Benford, S., Fahlén, L.E.: A spatial model of interaction in large virtual environments. In: ECSCW, p. 107 (1993)

27. Barkhuus, L., Dey, A.: Is context-aware computing taking control away from the user? three levels of interactivity examined. In: Dey, A.K., Schmidt, A., McCarthy, J.F. (eds.) UbiComp 2003. LNCS, vol. 2864, pp. 149–156. Springer, Heidelberg (2003)

28. Schmidt, A.: Implicit human computer interaction through context. Personal and Ubiquitous Computing 4(2-3), 191–199 (2006)

29. für Datenschutz Schleswig-Holstein und Institut für Wirtschaftsinformatik der Humboldt-Universität zu Berlin, U.L.: TAUCIS – Technikfolgen-Abschätzung Ubiquitäres Computing und Informationelle Selbstbestimmung (2006), http://www.datenschutzzentrum.de/taucis/index.htm

30. Barwise, J.: Situations and Attitudes. MIT-Bradford, Cambridge (1983)
31. Heckmann, D.: Distributed user modeling for situated interaction. GI Jahresta-gung (1), 266–270 (2005)
32. Heckmann, D., Schwartz, T., Brandherm, B., Schmitz, M., von Wilamowitz-Moellendorff, M.: Gumo - the general user model ontology. In: User Modeling, pp. 428–432 (2005)
33. Bryson, J.J.: Where should complexity go? Cooperation in complex agents with minimal communication. In: Truszkowski, W., Rouff, C., Hinchey, M. (eds.) Innovative Concepts for Agent-Based Systems, pp. 298–313. Springer, Heidelberg (2003)
34. Bryson, J.J., Stein, L.A.: Modularity and design in reactive intelligence. In: Proceedings of the 17th International Joint Conference on Artificial Intelligence, Seattle, pp. 1115–1120. Morgan Kaufmann, San Francisco (2001)
35. Weibelzahl, S.: Evaluation of adaptive systems, phd disertation (2003)

Application Challenges for Ubiquitous Knowledge Discovery

Koen Vanhoof[1] and Ina Lauth[2]

[1] Universiteit Hasselt
Department of Applied Economic Sciences
Research Group Data Analysis & Modelling,
Diepenbeek, Belgium
[2] Fraunhofer Institut Intelligente Analyse- und Informationssysteme,
Department Knowledge Discovery,
Sankt Augustin, Germany
koen.vanhoof@uhasselt.be, codrina.lauth@iais.fraunhofer.de

Very often new needs and the requirements of innovative applications act as powerful triggers for scientific advances. Today, the exponential diffusion of low-cost computing devices and the concept of intelligence, coupled with the easiness of generating, collecting and storing data, urges the development of methods and tools for managing, retrieving and exploiting information, anywhere and anytime.

This chapter, devoted to the applications of KDubiq, aims at:

– Providing an overview of potential applications in different areas, such as transportation, marketing, production, health care, logistics, bioinformatics, traffic safety, and multimedia.
– Showing that KDubiq technologies are necessary for future systems/solutions in these domains.

We give here a summary of a KDubiq challenges/requirements portfolio for different future applications. Concepts used in these portfolio have emerged in the previous chapters of this document. Although the overview is by no means exhaustive, it demonstrates that KDubiq technologies can be applied in a large number of domains. In some cases, prototypes are already existing, in other cases the application of data mining and machine learning techniques has not been attempted so far.

This chapter is based on discussions and joint work of the KDubiq Application working group. The main contributors for the application scenarios are credited in the footnotes to the individual sections. The texts have been edited and shortened to fit into the format of this chapter.

1 Mobility

Problem setting. The transport network is the backbone of economic, commercial, social and cultural activities in Europe. Individuals and economic agents are involved in various activities that require in some way the transportation

M. May and L. Saitta (Eds.): Ubiquitous Knowledge Discovery, LNAI 6202, pp. 108–125, 2010.
© Springer-Verlag Berlin Heidelberg 2010

of freights and/or people: commercial products have to be transported from factory to consumers, children participate in social and educational activities, adults travel to work or for leisure, tourists visit cultural sites, and so on. Due to personal and social restrictions, activities should be performed in a limited time frame, and some activities have to be finished before others can start. Practically every activity requires the use of the common infrastructure network. The demand for activities and the present environmental and time conditions determine how activities will be scheduled in space and time. The set of scheduled activities should then be brought in line with the supply of transport modes. Depending on the priorities set by the agent (minimizing the travel time or cost, maximizing the quantity transported, etc), a (set of) preferred transport mode(s) will be selected. Together with the current transport possibilities, this determines the modal split and consequently, the way the available transport infrastructure will be used.

Challenging problem: Traffic Congestion. One approach to relieve the pressure of traffic congestion is by means of advanced information technology which should be employed to timely and actively navigate vehicles so that limited transportation facilities can be sufficiently utilized [1]. There are examples of papers [2,3,4] that explore the possibility of establishing a VANET (Vehicular Ad hoc Network) to enable the dissemination of traffic and road conditions by independently moving vehicles. To achieve this goal, both sensors which are installed in the vehicle and information from other vehicles are used. VANETs are a ubiquitous data mining environment. Each vehicle receives data from various and unstable sources, such as in-vehicle sensors and V2V information, while it must be able to analyse this large amount of data in real-time and extract useful information for its own driver. Instead of a single point of analysis, VANETs distribute the data mining over all the vehicles in the ad hoc network.

Challenging problem: Simulating Multi-Modal Travel Behaviour. For changing from individual transport to public transport, information on park and ride facilities, on the departure of the public transportation vehicle and on the pedestrian route from the car to the station or platform is necessary. Returning from a public transport journey part requires assistance for retrieving the parked car. For changing within public transport, travellers have to know in which station to change, where the next public transport vehicle leaves, when it leaves and how to find the way from one platform to the other one. Changing from public transport to a footpath requires finding the right (nearest) exit of the station and the shortest footpath to the destination address. For changing from footpath to public transport, the calculation of the nearest station, which is connected to the destination station, is necessary. Moreover, the system has to assist with public transport information, the departure time of the next vehicle and orientation on the pedestrian route to the nearest station [5,6]. Systems that can offer this kind of information must be able to withdraw its data from various data sources, such as public transport schedules and geographical information systems. However, because the end-user is always on the run, the available networks and data sources are uncertain and constantly changing. Furthermore, this type of system must be implemented on small factor devices with mostly limited computational power, such as a PDA. Therefore, KDubiq technologies are necessary for such systems, because KDubiq techniques typically addresses these challenges like limited user interface, data streaming and resource-aware data mining. Existing projects that explore this idea are the OPEN-SPIRIT, the IM@GINE-IT and the MORYNE projects.[1]

The activity recognition prototype described in chapter 1 gives technical details how such a system could be implemented.

Challenging problem: Intelligent Route Planning. Classical route planning systems use information from the user about origin and destination and recently, through TMC signals, can also take into account current traffic conditions for calculating the most optimal route. Yet, this planning does not take into account the expected network load at a certain point in time on a particular location during the trip. Indeed, although a particular part of the network may be congested at the moment of planning a particular trip, it may no longer be congested at the moment of passing the particular area during the trip. As a result, the planned route at time t may be optimal given information about the state of the network at time t, but it may be suboptimal from an overall point of view because no information about the future state of the network is available. However, obtaining information about the future state of the network is not easy. In fact, one needs to possess information about other road users that will be sharing the same part of the network, i.e. their scheduled trips. In other words, to schedule routes in a more optimal way, information is needed about trips that other road users are expected to undertake in the near future (next hours). The exchange of information of this sort clearly provides challenges for ubiquitous technology. There is a need for a system, i.e. a communication network, processing capacity and intelligence to bring together information from different road users and to feed a feedback and control loop of information exchange. This challenge needs local and global learning techniques that are adaptive and continuous.

2   Traffic Safety

Problem setting. Each year there are some 50,000 fatalities on EU member states roads with approximately 2 million casualties. The European Road Safety Action Programme considers technology as an important element in increasing traffic safety. It is claimed that "The early detection of abnormal traffic conditions and the transmission of relevant data to drivers will make a significant contribution to improving road safety. The detection of abnormal traffic situations can be improved in the years to come by using vehicles themselves as sensors and by

---

[1] IM@GINE IT, Intelligent Mobility Agents, Advanced Positioning and Mapping Technologies INtEgration Interoperable MulTimodal location based services, http://www.imagineit-eu.com/; MORYNE, enhancement of public transport efficiency through the use of mobile sensor networks, http://www.fp6-moryne.org

centralising data in road traffic control centres thanks to the variety of means of communication available" [7].

Challenging problem: Hidden-Pedestrian Warning System. A pedestrian in front of a vehicle, parked at the side of the road, and who's hidden from the view of an approaching vehicle, finds himself unknowingly in a dangerous situation. However, the combination of in-vehicle sensor systems and V2V communication could tackle the potential danger. The sensors of the parked vehicle could detect objects in front and from behind. These data streams deliver a continuous stream of data to the intelligence module. The first challenge of the intelligence module is to identify the object in front of him as a pedestrian and not e.g. as an ordinary signpost. The same holds for the object behind the vehicle. The intelligence module must be able to identify the object as an approaching vehicle and not e.g. as another parked vehicle. KDubiq technology which focuses on real-time analysis from various sensor data streams could help the intelligence module in this task. The second challenge is to evaluate the potential danger of the situation prior to informing the approaching vehicle of the location of the pedestrian. A parked vehicle that would broadcast a message every time a pedestrian is in front of him would only cause redundant network traffic. KDubiq technology could offer data mining algorithms that can work efficiently in this type of environment to perform some sort of threat analysis. The parked vehicle will only send information about the pedestrian if the threat of the situation is high enough. In this threat analysis, the parked vehicle could contact an off-site analysis centre, which performs analysis and generates knowledge from historical traffic safety data. This off-site analysis centre could offer specific information on the historical probability of a hidden-pedestrian-accident at the specific location. Based on its sensor data and the historical data, the parked vehicle decides whether or not to warn the approaching vehicle.

If the approaching vehicle receives the information about the location of the pedestrian through V2V communication, it can combine this data stream with its own sensor data stream. If the pedestrian appears to be hidden from its own sensors, the vehicle could warn the driver about the hidden passenger. If the approaching vehicles also notices the pedestrian, it could evaluate the driver's behaviour to determine which kind of warning is appropriate (e.g. visual warning, sound warning, no warning). KDubiq offers technologies that can handle the constantly varying data sources. In this example, the V2V data streams may come and go as the approaching vehicle passes other cars. Furthermore, KDubiq also focuses on human-machine interfaces, which needs to be adapted to the situation. If the vehicle warns the driver for all pedestrians on the road, the driver is more likely to be distracted.

However, the approaching vehicle can only receive information about the hidden passenger if the parked vehicle has such a KDubiq intelligence module. Therefore, as an additional data source, the approaching vehicle could also receive V2I data from e.g. an installed camera nearby a school exit. This camera also possesses a little bit of KDubiq intelligence, enabling it to take over the task of the parked vehicle.

3   Healthcare

Problem setting. Ageing is a major challenge for European societies. Active ageing, with the help of individual living service (ILS), presents a major opportunity to harness technological progress for individual autonomy and dignity, for social inclusion and for establishing an effective and efficient health system attuned to the challenges of the next decades. ICT as a tool can provide complementary support, give new opportunities, like homecare and support to mobility, and remove the social or geographic distances between elderly people and their families. It can also reinforce older people's involvement in the community through the development of new activities, and through new ways of becoming part of human networks. By identifying and learning the needs of elderly people, KDubiq will create effective ways in which ICT can be integrated into their lives in order to provide the best possible support for their health and well being.

A wide range of existing and emerging devices for independent living can support these needs. Assistive devices from simple alarm bells to fully equipped smart rooms, telemedicine solutions and prevention of risks in daily activities through re-design of everyday tools can improve the autonomy of patients and the care they receive.

Challenging problem: Smart Homes. Smart homes are being used today in Europe and in other countries. The AID House, Gloucester Smart House for people with dementia and Dementia Friendly House (UK), Smart model house (NL), comHOME (Sweden), PROSAFE, Gardien, Grenoble HSH, Smardep (France) are some major examples of smart homes in Europe. The G.A.R.D.I.E.N. System in Grenoble is a successful application of a hospital smart room. In this hospital smart room, 8 infrared sensors are linked to a remote computer for collecting data in continuous mode. Other examples from North America are Rochester's 'Smart Medical Home', Georgia Tech's Aware Home (USA), MIT's House of the Future, Sherbrooke's Smart Home (Canada). 'Smart House in Tokushima' (Japan) is an example from Asia [8].

The MavHome project [9] is good example for a smart home that employs machine learning and data mining techniques for summarizing and predicting user activities. The key strength of their work is that the model does not require a human to create the model. A minimal amount of knowledge is required to automate and adapt – namely the automatable actions. The model is also not state restricted since not all possible states are considered but just the states actually observed. The data-driven model is only as good as the data that is used to generate it. The less consistent the inhabitant, the less ability there is to automate their life. The immediate goals are focused on techniques to assist with resource consumption reduction, to handle sensor noise, to provide a more natural partnership between the inhabitant and the environment. Researchers are also learning how to better design sensor networks and environments to promote more easily learnable (more consistent) patterns, developing a set of general design principles for building intelligent environments. The challenges are typical KDubiq challenges as resource-aware data mining, user interfaces, sensor networks, adaptation and continuous learning, and privacy.

4 Large Sensor Networks of Electrical Data[2]

Problem settings. Electricity distribution companies usually set their management operators on SCADA/DMS products (Supervisory Control and Data Acquisition / Distribution Management Systems). One of their important tasks is to forecast the electrical load (electricity demand) for a given sub-network of consumers. In SCADA/DMS systems, the load forecast functionality has to estimate for different time horizons, certain types of measures which are representative of system's load: active power, reactive power and current intensity. In the context of load forecast, near future is usually defined in the range of next hours to the limit of seven days, for what is called *short-term* load forecast.

Load forecast is a relevant auxiliary tool for operational management of an electricity distribution network, since it enables the identification of critical points in load evolution, allowing necessary corrections within available time. Companies make decisions to buy or sell energy based on load profiles and load forecast. In this context, several relevant learning tasks appear. For example, *monitoring* the grid requires the identification of failures, abnormal activities, extreme values and peaks, changes in the behavior of sensors or groups of sensors, etc. *Clustering* techniques, providing compact description of groups of sensors, are used to define users profiles (urban, rural, industrial, etc). Predictive models are used to predict the value measured by each sensor for different time horizons and prediction of peaks on the demand.

Our focus here is in electrical networks. The problem setting is much more general because the same problem definition (and solutions) can be easily applied to other domains, such as water or gas distribution.

Challenging problem: Load Forecast. This problem has been studied long time ago. The definition of the learning problem is quite well studied in terms of the set of relevant attributes. Several learning techniques have been used for load forecast: neural networks, Kalman filters, polynomials, ARIMA models, Wavelets, etc. In this scenario, the standard approach of learning static models trained from finite samples is not applicable. The electrical grid evolves over time, and static models quickly become outdated. The solution of retraining the learned models involves high costs. Techniques for online monitoring, online clustering sensors, and online prediction have been successfully developed and applied. However, most of the learning techniques were used in a centralized scenario were data is collected in a server.

This setting requires decision models that evolve over time. Learning techniques should be able to incorporate new information at the speed it is available,

---

[2] Contributed by João Gama and Pedro Pereira Rodrigues, LIAAD - INESC Porto, University of Porto, edited by K. Vanhoof.

should be able to adapt the decision models, detect changes and adapt the decision models to the most recent data. The most advanced techniques adapt the learning models in real-time.

This setting reveals a large and distributed sensor network. The thousands of distributed streams evolve over time, representing an interesting scenario for ubiquitous computing. This problem shares some of the requirements and objectives usually inherent to ubiquitous computing. Sensors are most of the times limited in resources such as memory and computational power, and communication between them is easily narrowed due to distance and hardware limitations. Moreover, given the limited resources and fast production of data, information has to be processed in quasi-real-time.

The challenging problems are the ability to deal with the large and variable number of sensors, producing high-speed data flows, the adaptation of the trained model to the dynamic behavior of the electrical demand. Real-time analysis might be done *in situ* using local information, minimizing communication costs.

The sensor networks problem we have described can easily include thousands of sensors, generating data at extremely fast rate. Furthermore, the electrical network spreads out geographically. The topology of the network and the position of the electrical-measuring sensors are known. From the geo-spatial information we can infer constraints in the admissible values of the electrical measures. Also the geo-spatial information can be used by the sensors themselves.

From a data mining perspective, this sensor network problem is characterized by a large number of variables (sensors), producing a continuous flow of data, eventually at high-speed, in a noisy and non-stationary environment. The number of sensors are variable, can increase or decrease by changes in the grid network topology.

With the evolution of hardware components, these sensors are acquiring computational power. The challenge will be to run the predictive model in the sensor itself. The sensors constitute a very large, distributed database where we can query not only the observed values but also the predicted values. All monitoring tasks run in the sensors themselves. Detection of abnormal behaviours, failures, complex correlation are done *in situ* using local information. Sensors would become smart devices, although with limited computational power, could detect neighbors and communicate with them. Data mining in this context becomes ubiquitous and distributed.

5 Organizing Networked Electronic Media[3]

Problem settings. Today, almost everybody owns very large personal collections of media (pictures, music, video) so that the retrieval even within the private personal collections becomes demanding. Given structures like the standard ontology indicating genre, artist, album, and year (used, e.g., by iTunes) are not sufficient. Collections need to be structured according to the needs and the views of particular users. Assuming a global view of the world (as put forward by the

---

[3] Contributed by Katharina Morik, University of Dortmund, edited by K. Vanhoof.

Semantic Web) does not reflect the very subjective way in which users handle information. Annotating resources carefully costs too much time and is rewarded only indirectly and after quite a while. Hence, users demand automated support for structuring their collections according to their subjective views. Machine learning or knowledge discovery offer services here: users can annotate some of their pictures, songs, or videos and a learning algorithm classifies the remaining items from the collections accordingly. Alternatively, new structures can automatically be imposed on so far unstructured items. The user can then choose the structure that suits her needs most. Recent advances in the web 2.0 domain show that both tasks can be achieved best in a collaborative setting, thus by exploiting the annotations produced by other users. Doing so in a way that allows for near real-time interaction in arbitrary mobile scenarios is an important challenge for ubiquitous data and multimedia mining methods.

Challenging problem: Organizing Multimedia. When observing people how they structure their collection, it becomes clear that not even a personal classification scheme is sufficient. Users have a different view of their collection depending on the occasions in which they want to use it. For instance, for a party you retrieve rather different music than for a candle-light dinner. Also places are related to views of the collection. While driving you might prefer rather different music than while working. A good host might play music for a special guest which he himself doesn't like usually. Hence, there is not one general preference structure per person, but several preference structures of the same person. These aspects of structuring collections have been investigated by Flasch et al. [10].

Of course, some people share their views. Networks allow users to access information in a transparent, location-independent way. Where some people share their views, it might happen as well that one person's favourite songs for car-driving are best liked by another person while cleaning the house, where for car-driving this person wants different music. Hence, browsing through another private collection should not follow the annotations of the collection through which someone is browsing, but the annotations/classifiers/features of the person who is browsing. In this way, a user might look at different providers' collections "through his own glasses". This can be achieved by classification. In a ubiquitous domain, there are however strong constraints on the response times. Especially, features have to be extracted in near real time. This is a new challenge. There are virtually infinitely many features to be extracted, but for each learning task (each annotation or tag) only a certain set of features leads to a high learning performance. A feature set which is valid for all tasks is hard to find [11]. Hence, learning the appropriate feature set for a particular learning task has become a learning task in its own right (see [12]) for an evolutionary composition of feature extraction methods for audio data. Nemoz is a framework for studying this kind of networked collaborative music organization in ubiquitous domains (see also the discussion in the introduction, Section 6.2). In Nemoz, each user may create arbitrary, personal classification schemes to organize her music. For instance, some users structure their collection according to mood and situations, others according to genres, etc. Some of these structures may overlap, e.g., the blues genre may cover songs which are also covered by a personal concept "melancholic" of a structure describing mood. Nemoz supports the users in structuring their media objects while not forcing them to use the same set of concepts or annotations. If an ad hoc network has been established, peers support each other in structuring. This ad hoc network is a real KDubiq environment with local and global processing and with a trade-off between processing and communication.

To give a first example, nodes can share features describing the items. Sharing all available features is, however, not feasible on networks with poor bandwidth, as are most current ad hoc networks. Therefore, methods for dynamically filtering features are needed that allow to achieve a high accuracy while requiring only moderate communication costs and computation time. Mierswa et al. [13] describe such a method, based on Support Vector Machines. It allows to identify relevant other nodes efficiently by comparing them based on a weight vector.

To give a second example, users can share tags attached to items. Again, a major challenge is to find relevant tags and combinations of tags in real-time in a distributed environment. The LACE algorithms deliver sound results by combining tag structures in a combinatorial way. Queries to other nodes are reduced to the search for a sample of items and can be performed based on existing peer-to-peer technology.

By recommending features and tags structures to other users, emerging views on the underlying space of objects are established. This approach naturally leads to a social filtering of such views. If someone creates a (partial) taxonomy found useful by many other users, it is often copied. If several taxonomies equally fit a query, a well-distributed taxonomy is recommended with higher probability. This pushes high quality taxonomies and allows to filter random or non-sense taxonomies.

In summary, distributed organization of music will be an ideal test case for the future success of ubiquitous data mining. First, the management of multi-media data is a very hard task because corresponding semantic descriptions depend on highly social and personal factors. Second, media collections are usually inherently distributed. Third, multi-media data is stored and managed on a large variety of different devices with very different capabilities concerning network connection and computational power.

While there have been first steps in solving these problems, there are still many open issues. One, for instance, concerns privacy. Currently Nemoz allows only for a very limited form of privacy and data mining could possibly cause security problems. Therefore, ad hoc feature extraction methods will have to be combined with privacy preserving data mining. Another open challenge is dealing with low memory and computational capabilities of many current mobile devices. The algorithms used in Nemoz are still quite demanding concerning computational power. There is a clear need for approximative data mining algorithms that can be executed efficiently on mobile devices still leading to satisfying results.

The overall vision of ubiquitous multimedia organization is that users can fully profit from what other users achieved, independently of their current position, of the device they are using and of their network connection.

6   Crisis Management[4]

Problem settings. In crisis situations, decision makers must respond appropriately to different adverse situations. Effective mitigation of hazardous situations requires profound situation awareness, i.e. knowledge of the relevant events in the affected environment. However, often crucial events cannot be observed directly. Consequently, such "hidden" events must be inferred through the appropriate interpretation of information sources, such as sensors and humans. Such interpretation is known as *information fusion*.

However, defining models for information fusion in real world applications is not trivial. Typical approaches to situation assessment in crisis management take a rather centralized view, in which all information that is available is sent to the operators. However, these centralized approaches suffer from several problems. They cannot cope with the growing complexity of problems and they cannot fully exploit the potential of modern communication and sensory technology. Namely, in modern applications huge amounts of information are sent to a central office which can result in clogging of communication resources, while operators must make sense out of huge amounts of information. At the same time, the results of their information fusion processes must be reliable since they have a mission critical impact, as misleading assessments can lead to wrong action selection which in turn may have devastating consequences on the further course of events. Moreover, due to the recent advances of communication and sensing technology we can obtain large amounts of relevant information which simply cannot be processed manually within the required time constraints.

Challenging problem: Distributed Perception networks. Distributed Perception Networks (DPN) [14] are capable of efficiently acquiring and interpreting huge amounts of relevant information from the crisis area by using sensors and humans. In particular, humans can be viewed as useful and versatile sources of information. Contrary to the existing artificial sensory systems, humans are good interpreters and can recognize a rich spectrum of phenomena.

By using DPNs, the information processing is not only performed at the control room, but mostly near the information sources themselves. In this way the DPNs support quick response times and graceful degradation [15]. A DPN is a KDubiq environment with local and global processing forming together in real time a distributed world model, created in response of the information need. The model results from "glueing together" partial world models, each running in a distributed agent. A participating agent can be a sensor, human (with mobile phone), or a computer. If more information is needed, the DPN system automatically contacts humans by sending SMS-queries to their mobile phones.

---

[4] Contributed by Marinus Maris, TU Delft and UVA, edited by K. Vanhoof.

A typical query is "do you smell a toxic gas?". The person that receives the message can respond with "yes" or "no". The answer is automatically integrated into the DPN-reasoning system to support the analysis of a subset of possible hypotheses [14]. Note that the hypotheses are selected automatically as the result of the dynamic modeling process. In this manner, many hypotheses can be swiftly evaluated and for example a leakage of a toxic gas is quickly confirmed. Reasoning is performed with Bayesian networks using probability calculation.

The challenge for DPN-type systems is to automatically send the appropriate questions to the persons that could give useful information and to evaluate a relevant number of hypotheses.

In general, the approach is relevant for applications where many relevant heterogeneous information sources have to be discovered and integrated into a fusion system at runtime. This is because the decision makers must be supplied only with relevant information distilled from large quantities of noisy and heterogeneous information. The seamless integration of information both from sensors and humans, as supported by DPN, suggests that this approach may be useful in a significant class of crisis management applications, such as detection and monitoring of toxic gases, forest fires, tsunamis, adverse meteorological phenomena, etc. A further big machine learning challenge would be to learn and to adapt the *structure* of the Bayesian network over time.

7   Process Monitoring[5]

Problem settings. Process monitoring is the task of assessing process behaviours based on the analysis of how observations of this process are evolving. A complete monitoring system involves not only the representation of system observations but also fault detection and diagnosis (location and identification) tasks. An appropriate definition of normal operation conditions (model) is needed for the system to be monitored in order to succeed in these tasks. Fault detection is accomplished by checking the consistency of observations with that model whereas fault diagnosis follows fault detection and takes profit of relations among variables defined in the model to isolate (locate) the origin of the fault and to identify its magnitude.

A variety of strategies have been proposed in the literature for fault detection and diagnosis. A complete review of such methods can be consulted in [16,17,18]. Since the basic principle of fault detection resides on the consistency between observations and models describing normal operation conditions, the quality of both models and observations is fundamental. According to the principles used to obtain such models one can distinguish between physics based models (those obtained from physical laws and basic engineering principles) and data-based models (those obtained from existing data).

In this way, data-driven methods have been introduced in the industry to improve monitoring systems following data mining schemas as external packages

---

[5] Contributed by Joaquim Melendez and Joan Colomer, Unversitat de Girona, edited by K. Vanhoof.

accessing those data bases. The most extended tools are those based on statistical principles and known as Statistical Process Control (SPC) methods.

Principal limitations of the majority of previous techniques reside in the numerical nature of data equally sampled and typically available in fixed length periods. The development of new data acquisition technologies (smart sensors, computer vision, PDAs, etc.) and its proliferation in the industry will require new paradigms to deal with this heterogeneity of data all together.

The irruption of ubiquitous technologies in the industry is evident and will change the way of sensing and monitoring the systems. The classical structure of a monitoring system centred on the process, where sensors are part of it is going to be complemented with the product point of view.

Challenging Problems: Performance Management. The benefit of monitoring processes and assets to improve efficiencies, reduce maintenance costs and to avoid un-planned outages and equipment failures is widely accepted and has driven most industries to make large investments in instrumentation and data management. Industry has come to a crossroads in its efforts to gain further benefits from investments in plant automation. For most "wired" plants today, the problem is no longer collecting the data but making sense of it in a time frame that supports decision-making. The problem of data overload is very real. Ubiquitous sensor networks (USN), constituted by sets of micro-sensors, or instruments, provided with computation and radio-communication capabilities and power autonomy is an example of this data overload. USN will flow within the product collecting its evolution in the transformation process and sensing the environment. USN will evolve with the product and the production necessities interacting with controllers and supervisors giving its particular view of the product evolution. Thus, the individual history of every component, or raw products, and its interactions will be available at anytime during the production.

Performance monitoring involves comparing actual measured performance to some design or expected level. When conditions slowly degrade over time, simple trend analysis can be used to raise alerts to plant operators that attention is required. For example, in power plants, system temperatures, pressures and flows can be monitored and the thermal performance can be computed from these measurements. This can be compared to design conditions and negative trends which develop over time are indicative of fouling or other performance related problems. The more difficult challenge is to identify when imminent equipment or component failure will cause an unplanned outage or will otherwise produce a change in plant performance. In some cases, simply trending the right parameter may be effective in avoiding this scenario, but usually degradation is due to a combination of several factors which cannot be predicted a-priori or easily detected from a casual review of the trend data.

Before analyzing the relationships between individual measurements, filtering and data smoothing techniques may be required upfront. In addition, heuristics can play an important role in preparing the data for analysis. The system employs several "smart filtering" approaches to minimize false alarms, to enhance signal to noise (S/N) and to reduce the effect of different time constants in the collection and transmission of data. The system can identify patterns in the data using both supervised and un-supervised learning approaches. Although models are a powerful approach to detect abnormal conditions, the underlying assumptions in the models may be invalid or change in real-time such that the simulations are no longer valid. Furthermore, the system may be too complex to model. A distributed KDubiq approach is needed that can model locally and globally.

At a low level, mining is necessary to manage the data load. Filtering, averaging, removing outliers, and other manipulation techniques let you create a composite view of physical phenomena by blending the readings from clusters of sensors. At a high level, decision-makers are less interested in sensor readings than they are in trying to answer management questions such as, "How should I reconfigure my process or allocate resources to respond to current conditions?" Just as data mining seeks to create useful information from structured business data, process mining seeks to create usable insight from sensor data.

A Vision of the Future. Continuous evolution of systems and flexible manufacturing needs the same flexibility to build and adapt models according to the process. Thus, model validation and exploitation techniques will evolve in the next years to take advantage of adaptive and evolutive models and providing mechanisms to reuse those models from one process to others. Caducity of models and prediction of its period of validity will be necessary in monitoring tasks to guarantee false alarm and miss detection ratios. Redundancy mechanisms needed for monitoring have to be defined in order to guarantee availability of consistent data for both model generation and on line monitoring. Detectability and diagnosticability indices will have to be calculated continuously to inform about monitoring performances through time. All these research challenges are directly related to the algorithmic issues discussed in Chapter 3.

8   Shopping Assistance[6]

Problem settings. The task of shopping is a very interesting scenario in which several aspects of KDubiq and ubiquitous computing in general can be combined. The basic challenge is to help the user to efficiently fulfil his shopping needs and to enhance his shopping experience. RFID will replace the nowadays often-used barcode on the short term. Apart from logistics, this RFID tagging of products can also be of benefit for the end-consumer if additional RFID sensors are installed in the shelves of the store, in the shopping cart of the consumer or even in the personal digital device of the user (e.g. their mobile phone or PDA). If the readings of the different RFID sensors are made available through an infrastructure in such a way that the user's device can collect these readings and combine them with additional personal information, a KDubiq environment is created and a variety of helpful applications become possible.

---

[6] Contributed by Tim Schwartz, University of Saarland and DFKI, Saarbrücken, edited by K. Vanhoof.

Challenging Problems: Mobile Shopping Assistant. A mobile shopping assistant can help the customer to browse through the contents of a shelf by showing additional product information, by automatically comparing two or more products of the same category or even by showing products that are currently out of stock but could nonetheless be of interest to the user. In this scenario, the input modality plays an important role: the desired product or products have to be indicated and commands have to be given (e.g. compare two products, or show the price of a product). By using the readings of the shelf's RFID sensor, the customer's PDA can keep track of which product has been removed from the shelf. If a product disappears from the shelf and does not appear in the cart, the system can infer that the customer is holding the said product. In combination with a user model of the customer [19], the system can interpret this behaviour in two ways: If the user model claims that the customer's experience with this kind of product is very sparse, the system infers that the user wants some additional information and it prints it on the screen. If the user is an expert and is holding the product for some time without putting it back in the shelf or in his cart, the system infers that the user may be searching for similar products and it displays a list of products of the same category. Since in this scenario the user is not carrying a PDA he is also able to pick two products at the same time. If the system recognizes this action and both products belong into the same category (e.g. flour), it infers that the customer wants to compare both products and displays a comparison chart. If the system detects that the user has put some products in their shopping cart, a plan recognition using a recipe book tries to infer if the user is planning to cook a certain dish and displays a list of products that may be useful for preparing this meal. If the user model contains the information that the customer is a vegetarian, the system will automatically omit any dishes that contain meat or fish. So on the PDA data mining models are used and updated.

Besides the products that a customer is interested in, other type of information can be very valuable in a shopping scenario: The current position of a user. To determine the location of a user in a building, active RFID tags and infrared beacons can be used. These tags and beacons are installed inside the building and contain or send out information about their own position. The user's PDA is enhanced with an active RFID read card, so that it can read the tags and receive the infrared beacons (with the built in infrared port). The PDA collects this data, that is provided by the environment, and uses dynamic Bayesian Networks to reason over its current position and heading direction [20,21]. A shopping mall or an airport typically has several specialized stores, like electronics stores, clothing stores and grocery stores. Using the knowledge about the customer's position and map knowledge, a system would now be able to navigate a user from one shop to the other according to his shopping list. Since it can be rather inconvenient to look at the display of a PDA all the time, the navigation aid can also be given by public displays that are mounted on the walls. The user can choose to give away his position information to the infrastructure and a Personal Navigator service, which has knowledge about the locations of the public displays, will then provide navigation instructions on the display that is nearest to the user.

A special presentation manager is used to organize the contents of each display, so that different information can be displayed concurrently (navigation instructions and flight schedule, for example). In combination with the user model and the sensed shopping actions, these public displays can also provide customized advertisements [22]. Instead of statically mounted displays, a steerable projector can be used to project information on any appropriate surface. This projector can also be used to highlight a certain product in a shelf [23].

Also the shop owner can infer a lot of knowledge. Today market basket analysis is a popular data mining application. However a market basket has limited information of the buying process. The user models contain much more valuable information for understanding the user's needs and the buying behaviour. So the owner can optimize assortment, allocation, price and design decisions.

Where it is obvious that some of these helpful applications may violate the customer's privacy, it is also important to notice that it can likewise violate the shop owner's commercial interests. E.g. navigating the customer from one compartment to another on the shortest route may prevent the customer from spontaneously buying something. Both interests, the customer's and the shop owner's, must be extensively discussed and systems like these should be implemented in such a way that the customer can always choose or at least make a trade-off between giving away personal information and gaining valuable service.

The main KDubiq challenges are the user-interface, the resource-aware modeling on the PDA and privacy.

9   Concluding Remarks

A classical data mining system has three components: data, algorithms and interfaces and runs on a single device in a secured environment. A KDubiq system

Table 1. Categorization of challenges in the application domains

	Technologies	Algorithms	Data Types	Privacy & Security	HCI
Mobility	A	A	B	A	A
Traffic Safety	A	A	B	A	A
Healthcare	B	C	A	A	A
Sensor networks	B	C	A	A	A
Networked electronic Media	C	A	A	C	A
Crisis management	B	B	C	B	A
Process monitoring	A	B	A	C	B
Retail marketing	B	A	A	A	A

[A] key challenge
[B] challenge
[C] weak challenge

however has multiple cooperating/communicating devices and works real time in a less secured environment. Therefore technologies and privacy & security are new research components. The blueprint follows this structure. Table 1 summarizes the current challenges in the different domains.

10   KDubiq Challenges versus Application Systems

In this section we give an overview of the appearance of mentioned characteristics/challenges in different future systems (Table 2). From this overview we can

Table 2. Summary of application features

	Activity Recognition	Vanets	Intelligent routeplanning	Hidden-pedest. warning	Smart home	Body Area Networks	Music mining	Shopping assistant
Ad-hoc networks		✓	✓	✓			✓	
Peer-to-peer								
Grid								
Wireless network		✓						
Communication limits	✓	✓						
Spatio-temporal sources	✓	✓	✓	✓	✓	✓		✓
Real time	✓	✓	✓	✓	✓	✓	✓	✓
Embedded processing	✓	✓	✓	✓	✓	✓		
Resource-aware computing	✓	✓			✓	✓		✓
Incremental learning	✓	✓	✓	✓	✓	✓		
Streaming		✓	✓	✓	✓	✓		✓
Feature selection							✓	
Concept drift	✓					✓		
Open learning		✓	✓	✓	✓			
Self-diagnosis		✓	✓					
Local learning	✓					✓		
Collaborative learning		✓	✓	✓			✓	
Evaluation metrics		✓	✓				✓	
Model exchange	✓	✓		✓	✓		✓	
Granularity			✓					
Resource-aware data types								
Location-aware data types	✓	✓						✓
Overlapping/contradictory Tagging							✓	
Privacy	✓	✓	✓		✓			✓
Security			✓	✓		✓		
HCI	✓	✓	✓	✓	✓			✓

deduce the generic character of the challenges and consider therefore these challenge as a new research domain. The reader will notice that some challenges (for example feature selection) are not new at all. This is true, the topic is not new but the new context or the new environment transform the topic in a new challenge. For example, feature selection has been well studied but feature selection of streaming data with concept drift is a new challenge.

References

1. Zhou, Y., Lu, H.: Mobile transportation information service system: the architecture and its technical solution. In: Proceedings of the 8th International IEEE Conference on Intelligent Transportation Systems, Vienna, Austria (2005)
2. Little, T., Argawal, A.: An information propagation scheme for vanets. In: Proceedings of the 8th International IEEE Conference on Intelligent Transportation Systems, Vienna, Austria (2005)
3. Chen, A., Khorashadi, B., Chuah, C.N., Ghosal, D., Zhang, H.: Smoothing vehicular traffic flow using vehicular-based ad hoc networking & computing. In: Proceedings of the 8th International IEEE Conference on Intelligent Transportation Systems, Vienna, Austria (2005)
4. Zhou, M., Korhonen, A., Malmi, L., Kosonen, I., Luttinen, T.: Integration of gis-t with real-time traffic simulation system: An application framework. In: Proceedings of the Transportation Research Board (2005)
5. Rehrl, K., Leitinger, S., Bruntsch, S., Mentz, H.: Assisting orientation and guidance for multimodal travelers in situations of modal change. In: Proceedings of the 8th International IEEE Conference on Intelligent Transportation Systems, Vienna, Austria (2005)
6. Yamashita, S., Hasegawa, T.: Pedestrian navigation system using textured paving blocks and its experiments. In: Proceedings of the 8th International IEEE Conference on Intelligent Transportation Systems, Vienna, Austria (2005)
7. Commission, E.: Communication from the Commission on the European Road Safety Action Programme. Halving the number of road accident victims in the European Union by 2010: a shared responsibility. Technical report, European Commission (2003) COM(2003) 311 final.
8. Comyn, G., Olsson, S., Guenzier, R., Özcivelek, R., Zinnbauer, D., Cabrera, M.: User needs in ICT research for independent living, with a focus on health aspects. Technical report, JRC (2006), http://ftp.jrc.es/eur22352en.pdf
9. Cook, D., Youngblood, G.M., Das, S.K.: A multi-agent approach to controlling a smart environment. In: AI and Smart Homes, pp. 165–182. Springer, Heidelberg (2006)
10. Flasch, O., Kaspari, A., Morik, K., Wurst, M.: Aspect-based tagging for collaborative media organisation. In: Proceedings of the ECML/PKDD Workshop on Ubiquitous Knowledge Discovery for Users (2006)
11. Pohle, T., Pampalk, E., Widmer, G.: Evaluation of frequently used audio features for classification of music into perceptual categories. In: Proceedings of the Fourth International Workshop on Content-Based Multimedia Indexing (2005)
12. Mierswa, I., Morik, K.: Automatic feature extraction for classifying audio data. Machine Learning Journal 58, 127–149 (2005)

13. Mierswa, I., Wurst, M.: Efficient case based feature construction for heterogeneous learning tasks. In: Gama, J., Camacho, R., Brazdil, P.B., Jorge, A.M., Torgo, L. (eds.) ECML 2005. LNCS (LNAI), vol. 3720, pp. 641–648. Springer, Heidelberg (2005)
14. Maris, M., Pavlin, G.: Distributed perception networks for crisis management. In: Proceedings of ISCRAM, Newark, NJ, USA (2006)
15. Pavlin, G., Maris, M., Nunnink, J.: An agent-based approach to distributed data and information fusion. In: Proceedings of Intelligent Agent Technology, Peking, China (2004)
16. Venkatasubramanian, V., Rengaswamy, R., Kavuri, S.: A review of process fault detection and diagnosis. part i: Quantitative model-based methods. Computers and Chemical Engineering 27(3), 293–311 (2003)
17. Venkatasubramanian, V., Rengaswamy, R., Kavuri, S.: A review of process fault detection and diagnosis. part ii: Qualitative methods and search strategies. Computers and Chemical Engineering 27(3), 293–311 (2003)
18. Venkatasubramanian, V., Rengaswamy, R., Kavuri, S.: A review of process fault detection and diagnosis. part iii: Process history based methods. Computers and Chemical Engineering 27(3), 293–311 (2003)
19. Heckmann, D.: Ubiquitous User Modeling. Akademische Verlagsgesellschaft Aka GmbH, Berlin (2006)
20. Brandherm, B., Schwartz, T.: Geo referenced dynamic bayesian networks for user positioning on mobile systems. In: Strang, T., Linnhoff-Popien, C. (eds.) LoCA 2005. LNCS, vol. 3479, pp. 223–234. Springer, Heidelberg (2005)
21. Schwartz, T., Brandherm, B., Heckmann, D.: Calculation of the user-direction in an always best positioned mobile localization system. In: Proceedings of the International Workshop on Artificial Intelligence in Mobile Systems (AIMS 2005), Salzburg, Austria (2005)
22. Stahl, C., Baus, J., Brandherm, B., Schmitz, M., Schwartz, T.: Navigational- and shopping assistance on the basis of user interactions in intelligent environments. In: Proceedings of the IEE International Workshop on Intelligent Environments (IE 2005). University of Essex, Colchester (2005)
23. Butz, A., Schneider, M., Spassova, M.: Searchlight - a lightweight search function for pervasive environments. In: Second International Conference on Pervasive Computing (2004)

Part II

Case Studies

On-Line Learning: Where Are We So Far?

Antoine Cornuéjols

AgroParisTech
UMR 518, Dept. MMIP
16, rue Claude-Bernard
F-75231 Paris cedex 05, France
antoine.cornuejols@agroparistech.fr

Abstract. After a long period of neglect, on-line learning is re-emerging as an important topic in machine learning. On one hand, this is due to new applications involving data flows, the detection of, or adaption to, changing conditions and long-life learning. On the other hand, it is now apparent that the current statistical theory of learning, based on the independent and stationary distribution assumption, has reached its limits and must be completed or superseded to account for sequencing effects, and more generally, for the information carried by the evolution of the data generation process.

This chapter first presents the current, still predominant paradigm. It then underlines the deviations to this framework introduced by new on-line learning settings, and the associated challenges that they raise both for devising novel algorithms and for developing a satisfactory new theory of learning. It concludes with a brief description of a new learning concept, called tracking, which may hint as to what could come off as algorithms and theoretical questions from looking anew to this all-pervading situation: never to stop learning.

1 Introduction

While Machine Learning is still a young field, having approximately 50 years of existence, its history is now sufficiently long that some historical perspective and deep trends can be perceived. Thus, the first learning algorithms were all incremental. For instance, the perceptron, the Checker program, ARCH or the Candidate Elimination Algorithm for Version Space learning, to name but a few. Various reasons motivated this state of affair. One was that these programs were, in part at least, aimed at simulating human learning, which is mostly incremental in nature. Another was that the very limited available computing power, by today's standard, prevented the storage and processing of large data bases of learning examples. However, this rule was completely overturned in the 80s. A wealth of new algorithms were developed: viz. decision trees, feed-forward neural networks, support vector machines, grammatical inference systems, and many more. Almost all are "batch" learners, meaning that they learn from a single batch of examples, optimizing some inductive criterion over the *whole* training

M. May and L. Saitta (Eds.): Ubiquitous Knowledge Discovery, LNAI 6202, pp. 129–147, 2010.

set. If new training instances are made available, then the learning process must start all over again from scratch.

It is interesting to examine reasons for this complete about-turn from the previous period. Thus, section 2 in particular provides the fundamentals of the now standard setting. Recent years, however, have witnessed a renewal of interest for on-line learning. Sections 3 provides reasons for this and the issues that are raised in consequence. Then, section 4, describes essential issues in on-line learning. Each of these issues remains essentially to solve, both at a theoretical level but also at the engineering level of conceiving new learning algorithms. Section 5, describes a special on-line learning framework called *tracking*. We show how new ideas could be brought to play in order to provide both original theoretical tools and learning methods to solve this problem. We think this nicely underlines the range of issues at play as well as the type of new ideas that we could call upon. Finally, we conclude with an appeal to a new scientific outlook for learning.

2   The Standard Setting: One-Shot and i.i.d.

An agent learns when it interacts with the world, using percepts to make decisions and take actions, and then measuring its performance, without which it would not be able to sense in what direction it should modify its decision making process. In order to learn, one has to be able to compare situations, that is to measure similarities or to make generalizations. One central concern in the study of learning has focussed on generalization and on questions such as: which conditions allow one to generalize? how to perform generalization? how to evaluate the confidence in the result of generalization?

Most studies in inductive learning assume that the learning agent comes across random feature vectors $\mathbf{x}$ (called the "observables"), which are generated according to the following two-stage process. First, a random class e.g. $y \in \{-1, 1\}$ is selected using the *a priori* probabilities $\mathbf{p}_y$; then, the observed feature vector $\mathbf{x}$ is generated according to the class-conditional distribution $\mathbf{p}_{\mathcal{X}|y}$. The distribution over labelled patterns is thus given by $\mathbf{p}_{\mathcal{X}y} = \mathbf{p}_y \mathbf{p}_{\mathcal{X}|y} = \mathbf{p}_{\mathcal{X}} \mathbf{p}_{y|\mathcal{X}}$.

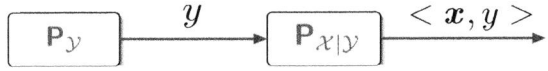

Fig. 1. The two-stage generation of learning examples

When acting as a *classifier*, the agent is facing the following problem: given a realization of the measured feature vector $\mathbf{x}$, decide whether the unknown object engendering $\mathbf{x}$ belongs to class -1 or 1. A classifier or decision rule, in this setting, is simply a map $h : \mathcal{X} \to \{-1, 1\}$, which determines the class $h(\mathbf{x})$ to which an observed feature vector $\mathbf{x}$ should be assigned. In the context of

Machine Learning, this map is called *a hypothesis*, hence the notation h[1]. It is thus possible to define the performance of a classifier (or hypothesis) as the *probability of error* given by:

$$L(h) = \mathbf{p}_{\mathcal{X}\mathcal{Y}}\{h(\mathbf{x}) \neq y\} \tag{1}$$

More generally, if different costs are assigned to different types of errors[2], specified through the definition of a *loss function* ℓ, defined as:

$$\begin{aligned}\ell(h) : \mathcal{X} \times \mathcal{Y} &\rightarrow \mathbb{R}_+ \\ (\mathbf{x}, y) &\mapsto \ell(h(\mathbf{x}), y)\end{aligned} \tag{2}$$

Then, the performance of a classifier is defined as a *risk*, which is an expectation over the possible events:

$$R(h) = \mathbb{E}[\ell(h(\mathbf{x}), y)] = \int_{\mathbf{x} \in \mathcal{X}, y \in \mathcal{Y}} \ell(h(\mathbf{x}), y)\, \mathbf{p}_{\mathcal{X}\mathcal{Y}}\, d(\mathbf{x}, y) \tag{3}$$

If the *a priori* probabilities $\mathbf{p}_y$ and conditional distributions $\mathbf{p}_{y|\mathcal{X}}$ are known, the optimal decision rule, in the sense of minimum probability of error (or of minimum risk) is the *Bayes decision rule*, denoted h^* and defined as:

$$h^*(\mathbf{x}) = \underset{y \in \{0,1\}}{\mathrm{ArgMin}}\big(\ell(y, 1-y)\, \mathbf{P}\{y|\mathbf{x}\}\big) \tag{4}$$

In many situations, however, these distributions are unknown or only partially known, but one is given a training set $\mathcal{S}_m = \{(\mathbf{x}_1, y_1), \ldots, (\mathbf{x}_m, y_m)\} \in (\mathcal{X} \times \mathcal{Y})^m$ supposed to be drawn according to the unknown probability distribution $\mathbf{p}_{\mathcal{X}\mathcal{Y}}$. The basic assumption enabling learning is that all the data (both observed and unseen) are generated by the same process, which is formalized by saying that the data is sampled independently from a fixed identical probability distribution (*i.i.d. sampling*).

The **learning problem** is thus: given a training set consisting of labelled objects supposedly drawn *i.i.d.* from the unknown distribution $\mathbf{p}_{\mathcal{X}\mathcal{Y}}$, find a function h that assigns labels to objects such that, if new objects are given, this function will label them correctly.

Short of attaining a perfect identification of the target dependency between the feature vectors and their label, the performance of a classifier or hypothesis is measured with the risk $R(h)$ (see equation 3). A large part of the theory in Machine Learning focuses on finding conditions for constructing good classifiers h whose risk is as close to $R^* = R(h^*)$ as possible.

A natural and simple approach is to consider a class $\mathcal{H}$ of hypotheses $h : \mathcal{X} \rightarrow \{-1, 1\}$ and to estimate the performance of the hypotheses based on their empirical performance measured on the learning set. The most obvious choice

---

[1] In Statistics, this notion is known as a *model*.
[2] In medicine, for instance, it is much more costly to miss an appendicitis diagnosis, than to decide to operate, only to discover it was a false alert.

to estimate the risk associated with a hypothesis is to measure its *empirical risk* on the learning set $\mathcal{S}_m$:

$$R_m(h) = \frac{1}{m} \sum_{i=1}^{m} \ell(h(\mathbf{x}_i), y_i) \tag{5}$$

which, in the case of binary classification with 0-1 loss, gives:

$$R_m(h) = \frac{1}{m} \sum_{i=1}^{m} \mathbb{I}_{(h(\mathbf{x}_i) \neq y_i)}, \tag{6}$$

where one counts the number of prediction errors on the training set.

In this framework, it is then natural to select a hypothesis with the minimal empirical risk as a most promising one to classify unseen events. This inductive criterion is called the *Empirical Risk Minimization* principle. According to it, the best candidate hypothesis is:

$$\hat{h}^* = \underset{h \in \mathcal{H}}{\text{ArgMin}}\, R_m(h) \tag{7}$$

However, the statistical theory of learning has shown that it is crucial that the hypothesis space from which candidate hypotheses are drawn be limited in terms of its expressive power. Specifically, one should not concentrate only on finding hypotheses that minimize the empirical risk irrespective of the hypothesis space, but one should indeed take into account its *capacity* or expressive power. In fact, the less diverse is $\mathcal{H}$, the tighter the link between the measured empirical risk $R_m(h)$ and the expected risk $R(h)$ for a given sample size m. This yields a regularized inductive criterion:

$$\hat{h}^* = \underset{h \in \mathcal{H}}{\text{ArgMin}} \left[R_m(h) + \text{Capacity}(\mathcal{H}) \right] \tag{8}$$

From this section, one can retain that *the inductive criterion*, be it empirical risk minimization, maximum likelihood or minimum description length principle, *plays an essential role in machine learning*. It both expresses the characteristics of the problem: misclassification costs and measures of performance, and allows us to analyze the conditions for successful induction. In addition, it did motivate several, if not most, modern learners (SVM, boosting, ...). However, it presupposes *a special kind of link between the past and the future*. The assumption is that both $\mathbf{p}_{\mathcal{X}}$ and $\mathbf{p}_{\mathcal{Y}|\mathcal{X}}$ are stationary, and that the data are drawn independently and identically from these probability distributions.

While these assumptions were seen as reasonable when most applications involved the analysis of limited static data bases, their validity appear increasingly questionable in face of new learning tasks. This is especially true for knowledge discovery from ubiquitous systems.

3   On-Line Learning: Motivations and Issues

Two new trends are shaping the future of machine learning. The first one is that *new types of data sources* are more and more taped in in order to discover regularities or tendencies that help understand the world and make decisions. This is in part due to the growing availability of digitalized observations series (e.g. environmental measurements over time) and to the birth of new sensor systems, that can both be spatially distributed and produce temporal data. Data are thus made increasingly available through unlimited streams that continuously flow, possibly at high speed. Furthermore, the underlying regularities may evolve over time rather than be stationary. Finally, the data is now often spatially as well as time situated. The second trend has to do with the learning systems themselves. They are indeed more and more embedded within complex data processing and management systems that are designed to be "long-life" systems. Consequently, the learning component itself must be able to process the data as it flows in while providing at any given time an hypothesis about the world. This entails that the learning systems must become incremental learning systems.

The overall upshot is that data can therefore no longer be considered i.i.d.. This forces the field to reconsider the basic and fundamental tenets of the theory of induction.

More precisely, the wealth of new applications and learning situations can be categorized into broad classes according to the underlying characteristics of the world.

1. Possibly **stationary world**
    - but the learner has *limited resources*. This is the case, for instance, of learning from very large data bases (e.g. Telecoms: millions of examples; EGEE systems in particle physics: billions of examples, ...)
    - but there are *"anytime" constraints* on learning that preclude to wait for all the necessary data to be observed and processed before hypotheses or decisions must be made. Data streaming is one typical example of this family of applications.
2. **The target concept is stationary** while **the distribution $p_{\mathcal{X}}$** of the descriptive, also known as explanatory, variables **is changing**. This is called "covariate shift" [12] and is currently the focus of some research effort. Active learning is naturally prone to covariate shift since the learning data, chosen by the learner, is not representative of the underlying distribution.
3. Finally, **the target dependencies $p_{y|\mathcal{X}}$ themselves might change over time**. This is also known as *concept drift*. This might occur because the world is changing (e.g. the customers's tastes change with fashion), or because learning is transfered from one task to another one. Another such situation occurs when learning is tutored with the help of a teacher who (carefully) chooses the sequence of learning tasks, with increasingly difficult rules to learn.

4   Aspects of On-Line Learning

4.1   Reducing Computational Cost

Since the advent of modern learning algorithms, in the 80s, the common wisdom has been that batch learning was to be preferred to on-line learning. One important reason was that on-line learning tends to be sensitive to the order of presentation of the training examples, which was considered as a nuisance. Furthermore, studies showed that batch gradient algorithms converge much more rapidly to the optimum $\hat{h}^\star$ of the empirical risk $R_m(\cdot)$ over a sample of size m, than the corresponding on-line learning algorithms.

More precisely, a *batch gradient algorithm* minimizes the empirical risk $R_m(h)$ using the following formula:

$$h_{k+1} \;=\; h_k - \gamma_k \, \nabla_h R_m(h_k) \;=\; h_k - \gamma_k \, \frac{1}{m} \sum_{i=1}^{m} \nabla_h \ell\big(h_k(\mathbf{x}_i), y_i\big) \qquad (9)$$

where the learning rate γ_k is a positive number. Studies have shown that $(h_k - \hat{h}^\star)^2$ converges like e^{-k}, where k denotes the kth epoch or sweep over the training set.

By contrast, an *on-line or stochastic gradient procedure* updates the current hypothesis on the basis of a single sample $(\mathbf{x}_t, y_t)$, usually picked randomly at each iteration.

$$h_{t+1} \;=\; h_t - \gamma_t \, \nabla_h \ell\big(h_t(\mathbf{x}_t), y_t\big) \qquad (10)$$

Under mild assumptions, on-line algorithms converge almost surely to a local minimum of the empirical risk. But, if they converge to the general area of the optimum at least as fast as batch algorithms, stochastic fluctuations due to the noisy gradient estimate make the hypothesis randomly wobble around the optimum region whose size decreases slowly. The analysis shows that the expectation $\mathbb{E}(h_t - \hat{h}^\star)^2$ converges like $1/t$ at best. This seems to condemn on-line algorithms.

However, this study begs two issues. The first one is that, in fact, one is not interested in the convergence to the exact minimum of the empirical risk, but rather in the convergence to the minimum expected risk R. Therefore, fine optimization is not required, and it is the one that is costly for on-line procedures. The second issue is that one should not only compare the convergence with respect to the number of learning steps, but one should also take into account the total computational costs involved. In this respect, on-line learning is quite simple to implement and only involves one random example at each iteration which can be discarded afterwards. On the contrary, each iteration of a batch algorithm involves a large summation over all the available examples, and memory must be allocated to hold these examples and to perform possibly complex computations, for instance if second order derivatives are estimated.

Studies [1,2] have shown that whereas a batch learning algorithm can process N learning examples, using the same computational resources an on-line learning algorithm can examine T instances, where T is of the order $\mathcal{O}(N \log \log N)$. Thus, for instance, while a batch learner could afford to examine 10,000 instances, a on-line learner could process $\approx 22,200$ instances, or about as twice as much!

This has profound implications for learning from large data sets or from data streams. Indeed, most of the time, learning is mainly limited by the fact that some informative examples have not yet been observed rather than by the fact that the examples already seen have not been fully exploited by the optimization process. When this is true, then **on-line algorithms may turn out to be vastly superior to their batch learning counterparts since, for the same computational resources they can process more examples**.

This is thus a first reason to seriously consider on-line learning, even when the world is stationary. Of course, this is even more so in face of changing conditions. But first, let us consider some issues raised in incremental learning irrespective of the conditions of the world.

4.2   Incremental Learning: Issues Raised

In principle, it is not difficult to produce an incremental learning system. It suffices to take a batch learner and to make it learn each time this is warranted on the basis of all the training examples seen so far. There are, however, at least two obstacles on the road to the actual implementation of such a scheme. First, this would require a memory size that would grow at least linearly with time (the growth could be worse if the computations required for learning are more than linear on the size of the training set). Additionally, the computation cost can be expected to grow at least in the same proportion. Second, this is ineffective in face of a changing environment because past data may become obsolete and harmful. Control of the memory is therefore required in incremental learning. The question is: how to carry out this control?

As an illustration, suppose we take a very simple incremental supervised learning system, the lazy learner that stores past examples and decides the label of a new unseen instance on the basis of the labels of its k nearest neighbors. After a while, it cannot afford to store all past data, and must per force select instances to be discarded. What should be the optimal forgetting strategy?

There are numerous options, including the following ones:

1. Discard the oldest training instance.
2. Discard the most obvious outlier in the training instances.
3. Discard the instance with the highest proportion of neighbors of the same label.
4. Discard a randomly chosen training instance.
5. Discard a training instance that is the farthest apart from instances of any other class.
6. ...

Each strategy is associated with an implicit model of the world and implies a computational load that may vary between "not worth to mention" (e.g. strategy 4) and "really worrisome" (e.g. strategies 2, 5). Most importantly, there is no best strategy. It all depends on the properties of the varying conditions of the world. Furthermore, as soon as forgetting is allowed to occur, the learning result is prone to be order dependent, that is to depend on the order in which the training instances have been considered [10].

It is worth mentioning that an important part of current research of data streaming systems centers on the question of which *summary* should be kept about past data (see for instance [4]).

To sum up, it was common wisdom that learning required carefully designed search strategies in order to find a (quasi) optimal hypothesis, or in order to chose the most informative examples in active learning. It now appears that **learning should, very generally, require forgetting** as well and that this entails a whole new search and optimization problem in its own right.

Control of the memory will soon emerge as an essential issue, both to limit the space and computational load, but also in order to adapt to the changing conditions of the world. In addition, **sequencing effects will have to be mastered**. It was usual to ignore or to try to reduce them, usually through some randomization process. It may become profitable to use them instead as ways to take into account the history of the environment.

4.3   Covariate Shift: Changing $\mathbf{p}_\mathcal{X}$

Even if the dependencies that link inputs to outputs are stationary, the distribution of the input pattern may vary over time, and, therefore, be different between learning and predicting, a situation called *covariate shift*. For instance, even though the fundamental characteristics of diseases are determined by biological rules and therefore are quite stable, their prevalence may present seasonal variations. Another common situation in learning arises when the distribution of learning instances is tweaked in order to facilitate learning. Thus, one may want to balance the classes of instances when some classes have few representatives, a situation commonly encountered in medical diagnosis. Similarly, in active learning, the training instances are selected by the learner, which, generally, leads to a distorted representation of the true underlying distribution. In these cases, the training data can no longer be considered as independent and distributed according to the true distribution $\mathbf{p}_\mathcal{X}$, and, therefore, the measured empirical risk is no longer an empirical measure of the true risk. Of course, all of this misrepresentation of the training data is ever more true with data streams corresponding to an evolving phenomenon.

In this case, it is known that the classical inductive criteria, such as (regularized) empirical risk minimization or the maximum likelihood estimator, lose their consistency: the learnt estimator or hypothesis does not converge in probability to the optimal one.

Suppose $\mathbf{p}_\mathcal{X}$ denotes the training distribution and $\mathbf{p}_{\mathcal{X}'}$ the test distribution. The performance of the hypothesis learnt using a training sample drawn according to $\mathbf{p}_\mathcal{X}$ depends on:

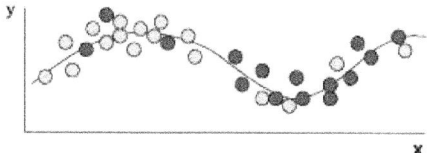

Fig. 2. In a stationary environment where the dependency $\mathbf{p}(y|\mathbf{x})$ is stationary, it may happen that the learning and the testing samples (respectively in light gray and dark gray) are not drawn from the same distribution (borrowed from [9])

- the performance of the hypothesis over $\mathbf{p}_\mathcal{X}$
- the similarity between $\mathbf{p}_\mathcal{X}$ and $\mathbf{p}_{\mathcal{X}'}$.

One obvious solution to regain consistency is to weight the training instances according to what is called their *importance*, that is the ratio of the test and training input densities: $\mathbf{p}_{\mathcal{X}'}(\mathbf{x})/\mathbf{p}_\mathcal{X}(\mathbf{x})$. One then gets the *importance weighted ERM* [12]:

$$\hat{h}^\star \;=\; \operatorname*{ArgMin}_{h \in \mathcal{H}} \left[\frac{1}{m} \sum_{i=1}^{m} \frac{\mathbf{p}_{\mathcal{X}'}(\mathbf{x})}{\mathbf{p}_\mathcal{X}(\mathbf{x})} \, \ell(h(\mathbf{x}_i), y_i) \right] \qquad (11)$$

Apart from stability considerations that imposes some modification, this new inductive criterion necessitates the estimation of the importance $\mathbf{p}_{\mathcal{X}'}(\mathbf{x})/\mathbf{p}_\mathcal{X}(\mathbf{x})$. However, the naive approach which is to first estimate the training and test input densities and then compute their ratio is rather impractical since density estimation is notoriously hard, especially in high dimensional cases. This is why recent research efforts have aimed at directly estimating the importance, bypassing density estimations. See e.g. [13] for more details.

4.4   Concept Drift: Changing $\mathbf{p}_{\mathcal{Y}|\mathcal{X}}$

Because of the increasing availability of data streams gathered over a long stretches of time, handling changing distributions and concept drifts has become a new important topic in machine learning. For instance, a company may find that her customer profile is varying over time. Likewise, in document filtering applications, the interests of the users may drift, or even abruptly change. The learning system should then revise and continuously adapt its model accordingly. We restrict ourselves here to the pattern recognition framework, that is to supervised learning of classes of patterns.

The learning problem may be characterized as follows. We suppose that data arrives in sequence, either one at a time or in small batches. Within each batch the data is independently and identically distributed with respect to a "local" distribution $\mathbf{p}_{\mathcal{X}\mathcal{Y}}(t)$. A concept drift occurs when the conditional distribution $\mathbf{p}_{\mathcal{Y}|\mathcal{X}}$ changes with time. The aim of the learner is to sequentially predict the label of the next example or batch, and to minimize the cumulated loss. Often this cumulated loss will correspond to the number of prediction errors.

It is assumed that the newly arrived data most closely resemble the current true concept. Furthermore, a reasonable assumption is that there exists some sort of temporal consistency in the changing environment corresponding to the fact that, most of the time, the underlying distribution of the data is changing continuously. These two assumptions imply that recent data carries more information about the current underlying concept than more ancient data.

Accordingly, learning in a changing environment is often handled by *keeping windows* of fixed or adaptive length on the data stream, or by *weighting data* or parts of the current hypothesis in accordance with their age or their utility for the learning task.

Whatever the approach, maintaining sliding windows or weights, the same tradeoff must be solved. On one hand, the system must be able to detect true variations against a noisy background, meaning it must be robust to irrelevant variations. On the other hand, the system should adapt as quickly as possible to variations in the environment in order to minimize its cumulated loss. Unfortunately, these two demands point to opposite strategies. Robustness to noise increases with the amount of data that is taken into account, but, in changing conditions, the oldest the data, the more likely it is obsolete. Therefore, at the same time, one would want to keep as much as possible information from the past, while reducing its importance for fear of being erroneously biased. Most works in concept drift have focussed on devising heuristics to solve this conundrum, that is **to control the memory of the past**.

Domingos and Hulten, [5], attack head on the issue of learning in face of very rapid data streams. They require their Very Fast Decision Tree learner to induce a decision tree on the fly so that the result is with high probability almost the same as the one that would be obtained with a batch learner but using only constant time and space complexity. For this, they rely on the statistical theory of learning, specifically on Hoeffding formula, in order to compute bounds on the required minimal training set size to approximate the optimal decision function within a given error factor. Here, this bound is iteratively used for the determination of each node in the tree. Of course, this learning method uses much more data than would be strictly necessary to ensure to get a good approximation of the target tree, but it is assumed that there is an over-supply of data. The interesting idea in this approach therefore lies in using the theory to get estimates of the required window size on the past sequence of data. The limit, however, is that it assumes stationarity. In a subsequent paper, [7], the problem of time-changing environment was tackled and another useful concept was put forward: to grow a new tree when the data is starting to drift away from the previous distribution and to start using the new tree when it becomes more accurate than the old one. This enables to use past information as long as it is useful, and to overcome to some degree an explicit trade-off for the choice of the window size.

Managing window size

If a fixed size is to be chosen for sliding windows, the choice results from a compromise between fast adaptability (small window to the risk of under-fitting) and good generalization (large window). It can only be made on the basis of assumptions about the pace of the changes in the environment. If no such well-informed assumption is possible, one has to rely on adaptive strategies for the window size management. The challenge in automatically adjusting the size of the window is to minimize at each time the expected loss on new examples. This requires that the model of the data is as accurate as possible at each time step. When the underlying distribution is stable, the window size can safely increase, enabling better generalization and the learning of more detailed and accurate models of the environment, whereas when the distribution is changing, the window size must shorten in order to not incorporate obsolete and harmful training instances (see figure 3).

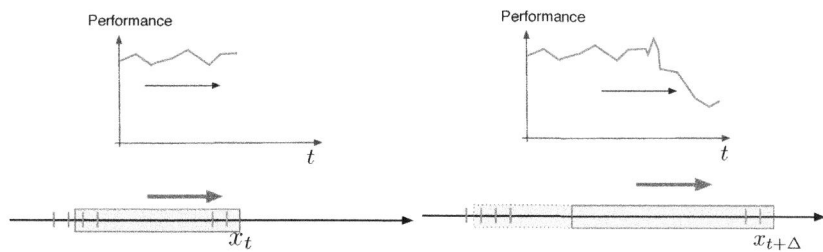

Fig. 3. When the underlying distribution is stable, or at least, the performance is, the size of the window may increas (left). When the performance is falling, betraying a change in the distribution, the window size must be reduced (right).

One way to implement an adaptive strategy goes as follows. Suppose that batches of equal size arrive at each time step. Suppose also that the current time step is T while the learning process started at time $t = 0$. A classifier is learnt over the most recent batch, and is tested over every preceding windows of size less than $\max(max\_length, T)$. The window of maximal size for which the error is $< \varepsilon$ for a given ε is kept. A classifier is learned over this window and is used for predicting the class of the newly arriving unclassified example(s) (see figure 3). In case of a abrupt change of distribution, it may happen that the learning window is reduced to the most recent batch of data.

One problem with this strategy is that there is no memory of the past when the underlying distribution has changed. Let us suppose, for instance, that the underlying distribution switches from distribution $\mathbf{p}^1_{\mathcal{Y}|\mathcal{X}}$ to distribution $\mathbf{p}^2_{\mathcal{Y}|\mathcal{X}}$, and then reverses back to distribution $\mathbf{p}^1_{\mathcal{Y}|\mathcal{X}}$. Then, when the first distribution $\mathbf{p}^1_{\mathcal{Y}|\mathcal{X}}$ rules the generation of data once more, learning will have to restart all over again. This is why other approaches have been proposed [11]. For instance, one may envision that all past batches for which the prediction error of the classifier learnt over the most recent batch is $< \varepsilon$ are kept for learning the classifier used for prediction (see figure 4). In this way, past data that seems relevant to the current learning situation may be used for learning. The relearning time may thus be greatly reduced, enabling better performance.

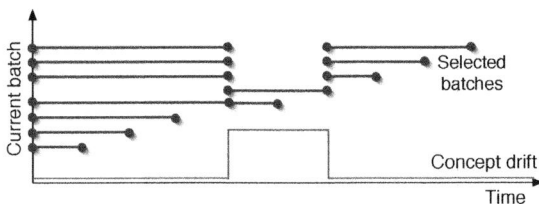

Fig. 4. Adaptive management of window size with memory if a concept is encountered again. (Borrowed from [11]).

Weighting past examples

There have been very numerous proposals for selecting or weighting past instances in order to confront changing environments. For the purpose of illustration, we mention here one technique called "locally weighted forgetting" and used in nearest neighbors classification. In this scheme, all data starts with weight 1, and when a new data point is observed, the weights of the k nearest neighbors are adjusted according to the following rule: (1) The closer the data to the new sample, the more the weight is decayed; (2) If weight drops below some threshold, remove data.

In this way, it is hoped that only sufficiently scattered representatives of the data are kept. But it is also possible to maintain weights on hypotheses rather than on data points.

Weighting past hypotheses

In the wake of the success of boosting techniques, ensemble methods for tracking concept drift have been recently proposed. The overall strategy is the following.

1. Learn a number of models on different parts of the data.
2. Weigh classifiers according to recent performance.
3. If classifier performance degrades, replace it by a new classifier.

More specifically, for instance, Kolter and Maloof in [8] describe the "dynamic weighted majority" algorithm that dynamically creates and removes weighted classifiers in response to changes in performance.

1. Classifiers in ensemble have initially a weight of 1
2. For each new instance:
   - if a classifier predicts incorrectly, reduce its weight;
   - if weight drops below threshold, remove classifier;
   - if the ensemble of current classifiers then predicts incorrectly, install new classifier;
   - finally, all classifiers are (incrementally) updated by considering new instance.

Among other proposals involving ensemble methods, like [15], one is standing out by introducing an intriguing and tempting idea [11]. Suppose that the underlying data distribution is continuously changing from concept 1 to concept 2, is it possible to learn concept 2 before the end of the transition? (See figure 5). Under some admittedly restrictive assumptions, this turns out indeed to be possible.

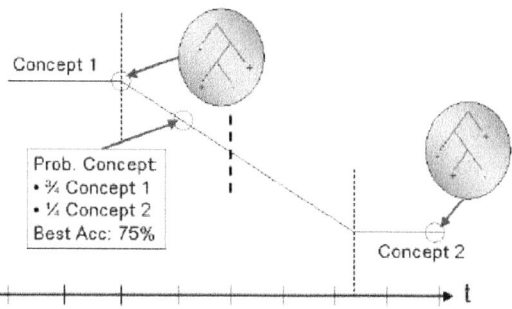

Fig. 5. Continuous concept drift changing linearly from concept 1 to concept 2. It is optimal to predict concept 1 before the dotted line, and concept 2 afterwards. (Borrowed from [11]).

The main assumption is that, during the concept drift, the training instances are sampled from a mixture distribution, that is a weighted combination of the two pure distributions characterizing concept 1 and concept 2. The optimal model during the concept drift can only be derived according to the Bayes's optimal rule. The idea is to decompose the mixture distribution as soon as the concept drift starts. It this respect, the boosting algorithm is very seducing since it is based on the principle to sample the data points at each step *orthogonally* to the current distribution. In the approach of Scholz and Klinkenberg, this becomes sampling orthogonally to the distribution induced by the prediction of h_1, the hypothesis learnt from data corresponding to concept 1. A careful analysis leads to a weighting scheme that modifies the weight of the examples with respect to hypothesis h_1 so that the new distribution reflects the characteristics of the new incoming distribution.

Lessons about concept drift
Learning in the presence of concept drift is still very much an open research issue even though a lot of interesting ideas and heuristics have been put forward. Overall, the existing methods are sometimes efficient in their respective application domains, but they usually require fine tuning. Furthermore, they are still not easily transferable to other domains. This denotes a lack of a satisfying theoretical ground.

There are indeed relatively few theoretical analyses, and most of them date back to the early 90s. One significant work is the one by Helmbold and Long [6].

They study the conditions for PAC learning with an error of ε in the presence of concept drift. This depends upon the diversity of the hypothesis space $\mathcal{H}$ and on the speed of the drift (measured as the probability that the 2 subsequent concepts disagree on a randomly drawn example). The outcome is a bound of the size of the required window size. Regrettably, these bounds are usually impractically large. This is due in part to the adversary protocol used in the analysis.

Besides the shortcoming of our current theoretical understanding, there are other desirable developments. Among them is the need for the capability to recognize and treat recurring contexts, for instance associated with seasonal variations, so that old models can be quickly recovered if appropriate. But, more significantly, there is a growing feeling that it would be profitable to **focus on the changes themselves** rather than merely trying to follow concept drifts as closely as possible. Reasoning about the "second derivatives" of the evolving situation and representing them would allow for quicker adaptation, as well as interpretability about what has changed and how. This raises the issue of having models for change and to incorporate them in new appropriate inductive criteria.

5   A New Perspective on On-Line Learning

In the following, we focus on a special case of learning task that has been conjured up in a recent paper by Sutton, Koop and Silver [14], called *tracking*.

5.1   The Tracking Problem

Sutton et al. argue that while most existing learning systems are geared to find a single best solution to the learning problem, one that applies to any possible input $\mathbf{x} \in \mathcal{X}$, it might be possible that better performance be attained with the same amount of training data and computing resources by tracking the current situation rather than by searching the best overall model of the world. In this view, the agent continuously learns a local model that applies to the situations that can be encountered at the time being.

More precisely, it is assumed that the learning agent encounters different parts of the environment at different times. The underlying distribution on $\mathcal{X}$ is now a function of time: $\mathbf{p}_{\mathcal{X}}(t)$, in such a manner that the data are not identically and independently distributed, but are governed by some time dependent process, like, for instance, a Markov decision process. In this case, it might be advantageous for the agent to adapt to the local environment defined by $\mathbf{p}_{\mathcal{X}}(t)$ and, possibly, by $\mathbf{p}_{\mathcal{Y}|\mathcal{X}}(t)$, if the later one is evolving too. Figure 6 schematizes the evolution of such a data-driven agent.

Temporal consistency, which we loosely define as the fact that $\mathbf{p}_{\mathcal{X}}(t)$ and $\mathbf{p}_{\mathcal{Y}|\mathcal{X}}(t)$ tend to evolve with cumulated bounded variation over limited periods of time, offers the opportunity to perform well in term of predictions by learning simple models with limited resources. Indeed, because of temporal consistency, the learner may expect that it will have to make predictions about inputs that lies in its "local environment". In addition, temporal consistency imposes that

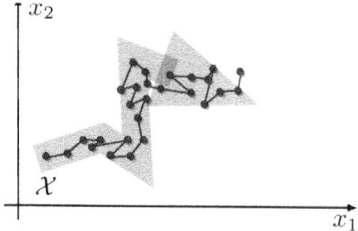

Fig. 6. In tracking, the learning agent receives input that are driven by a time dependent process. It therefore encounters different parts of the environment at different times.

the laws governing the local environment are simpler than the laws governing the whole input space and the whole time evolution of the world. Thus, even though the overall model of the world and it's time evolution may be arbitrary complex, it can be expected that, locally, in term of both input space and time, simple models may suffice for appropriate decisions.

Figure 7 illustrates this in a simple but extreme case. Suppose that $\mathcal{X} = \mathbb{R}$, and that the target dependency $\mathbf{p}_{y|\mathcal{X}}(t)$ is stationary and takes the form of a piecewise linear curve. Suppose that the local environment of the learning agent may be depicted by a window of limited size. As the agent is exploring $\mathcal{X}$, passively or actively, it may perform rather accurate predictions solely by maintaining a model of the world that is simply a "constant" prediction. This constant is updated with time, a process described by *tracking*.

Therefore, if time consistency holds, good prediction performance can be obtained with less computational cost than the classical batch learning. At each time step, the learner is searching for simpler models of the world and does so on the basis of limited amount of past data. There is thus a kind of spectrum to be expected along the following lines:

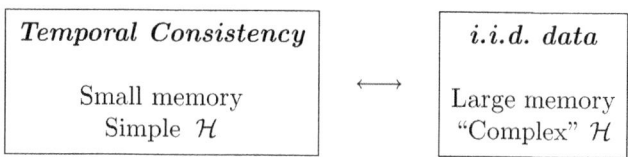

If, intuitively, a tracking strategy seems advantageous, several questions remain to be answered. The tracking problem, well-known in engineering sciences, but new in machine learning, needs to be formally defined. More importantly, we do not know yet how to measure the position of a learning problem along the afore-mentioned spectrum and how to evaluate the advantage in terms of learning resources needed for a given performance level in terms of prediction. In fact, the classical notion of expected performance and the associated risk formula certainly needs to be revisited. Finally, all of this, yet to come, analysis should be turned into new learning strategies.

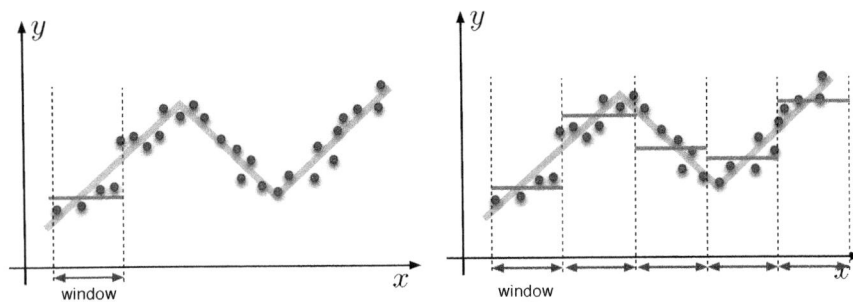

Fig. 7. Even though the world involves a piecewise linear law, the learning agent may perform well by maintaining a very simple model, a constant, over its local environment

In the following, we just propose a glimpse of the kind of new inductive criteria that could be interesting to consider.

5.2   A New Perspective for On-Line Induction

In the classical stationary framework, where learning is supposed to occur once for all from the available training data, the expected performance is defined with respect to each possible model of the world h:

$$R(h) \;=\; \mathbb{E}[\ell(h(\mathbf{x}), y)] \;=\; \int_{\mathbf{x} \in \mathcal{X}, y \in \mathcal{Y}} \ell(h(\mathbf{x}), y)\, \mathbf{p}_{\mathcal{X}\mathcal{Y}}\, d(\mathbf{x}, y)$$

which gives rise to the inductive criterion based on empirical risk:

$$R_m(h) \;=\; \frac{1}{m} \sum_{i=1}^{m} \ell(h(\mathbf{x}_i), y_i)$$

With on-line learning of successive hypotheses, these criteria can no longer be used as such. While it is clear that the performance still involves a cumulated loss over time, neither the model of the world h, nor the underlying data distribution $\mathbf{p}_{\mathcal{X}\mathcal{Y}}$ are stationary. It is therefore necessary to include their variations in the performance and inductive criteria. In fact, not only to include these variations, but to make them the focus of the optimization problem.

Indeed, the very notion of training sample needs to be cross-examined. In on-line learning, it is rarely the case that the learning system will be submitted twice to the same kind of *history*. Therefore, past data cannot be considered as representative of what will happen next. In other words, all past distributions $\mathbf{p}_{\mathcal{X}}(t - k)$, and, possibly as well, all past $\mathbf{p}_{\mathcal{Y}|\mathcal{X}}(t - k)$ for $0 \leq k \leq t$ may very well never be encountered again.

But, then, what would link the past to the future and allow for prediction and induction? One inductive assumption is that the underlying regularity to be learned is the rule that governs the variations of the environment. Denote r

the model of the rule that the learner tries to estimate, then the risk associated with such a rule can be expressed as:

$$R(r) = \mathbb{E}[\ell(h_t(\mathbf{x}_t), y_t)] = \int_t \ell(h_t(\mathbf{x}_t), y_t) \, \mathbf{p}_{\mathcal{XY}}(t) \, d(\mathbf{x}_t, y_t) \qquad (12)$$

While the rule r does not appear explicitly in the above expression, it is nonetheless present by the fact that the successive hypotheses h_t are linked by the rule:

$$h_{t+dt} = r(h_t, (\mathbf{x}_t, y_t), \text{memory}, \ell, dt) \qquad (13)$$

or, if time flows in discrete time steps:

$$h_{t+1} = r(h_t, (\mathbf{x}_t, y_t), \text{memory}, \ell) \qquad (14)$$

Here, the *memory* term is used to denote what trace of the past data is used by the update rule r.

Then, a possible inductive criterion, based on past data from time 0 to time T, could be of the form:

$$L_{\langle 0,T \rangle}(r) = \underbrace{\sum_{t=1}^{T} \ell(h_t(\mathbf{x}_t), y_t)}_{\text{classical cumulated loss}} + \underbrace{\lambda \sum_{t=1}^{T} ||h_t - h_{t-1}||^2 + \text{Capacity}(\mathcal{R})}_{\text{new criterion on } r} \qquad (15)$$

Where $\lambda > 0$ is a parameter weighting the importance of conditions over the regularity of the rule r. This regularity is conditioned both by the cumulated variations over h_t _temporal consistency imposes limited variations_, and by the complexity of the possible rules. Indeed, the *capacity* is a function of the *memory* used for updating the current hypothesis at each time step. This memory should be kept as limited as possible.

We do not delve into details within the limited scope of this chapter, but it is obvious that this kind of criterion is related to the theory of reinforcement learning and the underlying assumption of a Markov Decision Process. There also, what the learning agent is estimating are the rules that govern the transition from one state to the following and the reward function attached to states transitions[3].

6 Conclusions

Recent years have witnessed a wealth of emerging applications that can not be solved within the classical inductive setting. New learning tasks often involve data coming in unlimited streams and long-life learning systems that, in addition, have limited computational resources. The fact that data can no longer

---

[3] It must be noticed that the problem of the performance criterion is approached quite differently in the theory of on-line learning based on regret criteria. For lack of space, we defer the reader to [3].

be considered as identically and independently distributed, and that the learner needs, per force, to implement on-line learning raises important new issues and announces profound evolutions of the field of machine learning.

Among the list of open questions are the following ones:

- How to deal with non i.i.d. data?
  - What to memorize? / What to forget?
  - How to cope with or take advantage of *ordering effects*?
  - How to *facilitate future learning*, what should be transfered? Representations, learned rules, ...?
- What should the inductive criterion be?
  - How to take the *computational resources* into the inductive criterion?
  - What kind of regularity should we optimize: $h \in \mathcal{H}$ or $r \in \mathcal{R}$?

There is already a growing body of work that touches on these questions. *Covariate shift, transduction, concept drift, tracking, transfer of learning*, even *teachability*, are subject matters that bear on the issue of on-line learning.

One important clue seems to be that, in evolving environments, the changes themselves should be the focus of learning. Works in concept drift have shown that this can accelerate recovery of useful past regularity, but, more generally, the analysis of possible new inductive criteria adapted to the problem of on-line learning seem to point to that direction as well.

In any case, whatever will be its scientific outcome, the present time is a privileged one for machine learning, a time for exciting research both for a better fundamental understanding of learning and for the design of new learning techniques. Ubiquitous learning environments, especially, are both the fuel and the beneficiaries of these incoming developments.

References

1. Bottou, L., Bousquet, O.: The tradeoffs of large scale learning. In: Platt, J.C., Koller, D., Singer, Y., Roweis, S. (eds.) Advances in Neural Information Processing Systems, NIPS Foundation, vol. 20, pp. 161–168 (2008), http://books.nips.cc
2. Bottou, L., LeCun, Y.: On-line learning for very large datasets. Applied Stochastic Models in Business and Industry 21(2), 137–151 (2005)
3. Cesa-Bianchi, N., Lugosi, G.: Prediction, learning and games. Cambridge University Press, Cambridge (2006)
4. Cormode, G., Muthukrishnan, S.: An improved data stream summary: The count-min sketch and its applications. Journal of Algorithms 55(1), 58–75 (2005)
5. Domingos, P., Hulten, G.: Mining high-speed data streams. In: Proceedings of the Sixth International Conference on Knowledge Discovery and Data Mining, pp. 71–80. ACM Press, New York (2000)
6. Helmbold, D., Long, P.: Tracking drifting concepts by minimizing disagreements. Machine Learning 14(1), 27–45 (1994)
7. Hulten, G., Spencer, L., Domingos, P.: Mining time-changing data streams. In: KDD 2001: Proceedings of the Seventh ACM SIGKDD International Conference on Knowledge Discovery and Data Mining, pp. 97–106. ACM, New York (2001)

8. Kolter, J.Z., Maloof, M.A.: Dynamic weighted majority: An ensemble method for drifting concepts. Journal of Machine Learning Research 8, 2755–2790 (2007)
9. Quinonero-Candela, J., Sugiyama, M., Schwaighofer, A., Lawrence, N.: Dataset shift in machine learning. MIT Press, Cambridge (2009)
10. Ritter, F., Nerb, J., Lehtinen, E., O'Shea, T. (eds.): In order to learn. How the sequence of topics influences learning. Oxford University Press, Oxford (2007)
11. Scholz, M., Klinkenberg, R.: Boosting classifiers for drifting concepts. Intelligent Data Analysis 11(1), 3–28 (2007)
12. Sugiyama, M., Kraudelat, M., Müller, K.-R.: Covariate shift adaptation by importance weighted cross validation. Journal of Machine Learning Research 8, 985–1005 (2007)
13. Sugiyama, M., Nakajima, S., Kashima, H., Von Buenau, P., Kawanabe, M.: Direct importance estimation with model selection and its application to covariate shift adaptation. In: Platt, J.C., Koller, D., Singer, Y., Roweis, S.T. (eds.) NIPS. MIT Press, Cambridge (2007)
14. Sutton, R., Koop, A., Silver, D.: On the role of tracking in stationary environments. In: Proceedings of the 24th International Conference on Machine Learning, Corvalis, Oregon, pp. 871–878. ACM, New York (2007)
15. Wang, H., Fan, W., Yu, P.S., Han, J.: Mining concept-drifting data streams using ensemble classifiers. In: Proceedings of the Ninth International Conference on Knowledge Discovery and Data Mining (KDD 2003), pp. 226–235 (2003)

Change Detection with Kalman Filter and CUSUM

Milton Severo[2,4] and João Gama[1,3,4]

[1] Faculty of Economics
[2] Department of Hygiene and Epidemiology, Faculty of Medicine, University of Porto
[3] LIAAD - INESC Porto LA
[4] University of Porto, Portugal
milton@med.up.pt, jgama@fep.up.pt

Abstract. In most challenging applications learning algorithms act in dynamic environments where the data is collected over time. A desirable property of these algorithms is the ability of incremental incorporating new data in the actual decision model. Several incremental learning algorithms have been proposed. However most of them make the assumption that the examples are drawn from a stationary distribution [14]. The aim of this study is to present a detection system (DSKC) for regression problems. The system is modular and works as a post-processor of a regressor. It is composed by a regression predictor, a Kalman filter and a Cumulative Sum of Recursive Residual (CUSUM) change detector. The system continuously monitors the error of the regression model. A significant increase of the error is interpreted as a change in the distribution that generates the examples over time. When a change is detected, the actual regression model is deleted and a new one is constructed. In this paper we tested DSKC with a set of three artificial experiments, and two real-world datasets: a Physiological dataset and a clinic dataset of sleep apnoea. Sleep apnoea is a common disorder characterized by periods of breathing cessation (apnoea) and periods of reduced breathing (hypopnea) [7]. This is a real-world application where the goal is to detect changes in the signals that monitor breathing. The experimental results showed that the system detected changes fast and with high probability. The results also showed that the system is robust to false alarms and can be applied with efficiency to problems where the information is available over time.

1 Introduction

In most challenging applications learning algorithms act in dynamic environments where the data is collected over time. A desirable property of these algorithms is the ability of incremental incorporating new data in the actual decision model. Several incremental learning algorithms have been proposed to deal with this ability (e.g., [5,13,6]). However most learning algorithms, including the incremental ones, assume that the examples are drawn from a stationary distribution [14]. In this paper we study learning problems where the process

M. May and L. Saitta (Eds.): Ubiquitous Knowledge Discovery, LNAI 6202, pp. 148–162, 2010.
© Springer-Verlag Berlin Heidelberg 2010

generating data is not strictly stationary. In most of real world applications, the target concept could gradually change over time. The ability to incorporate this concept drift is a natural extension for incremental learning systems.

In many practical problems arising in quality control, signal processing, monitoring in industrial plants or biomedical, the target concept may change rapidly [2]. For this reason, it is essential to construct algorithms with the purpose of detecting changes in the target concept. If we can identify abrupt changes of target concept, we can re-learn the concept using only the relevant information. There are two types of approaches to this problem: methods where the learning algorithm includes the detection mechanism, and approaches where the detection mechanism is outside (working as a wrapper) of the learning algorithm. The second approach has the advantage of being independent of the learning algorithm used. There are also several methods for solving change detection problems: time windows, weighting examples according their utility or age, etc [10]. In the machine learning community few works address this problem. In [17] a method for structural break detection is presented. The method is an intensive-computing algorithm not applicable for processing large datasets.

The work presented here follows a time-window approach. Our focus is determining the appropriate size of the time window. We use a Kalman filter [15,20] that smoothes regression model residuals associated with a change detection CUSUM method [2,4,11]. The Kalman filter is widely used in aeronautics and engineering for two main purposes: for combining measurements of the same variables but from different sensors, and for combining an inexact forecast of system's state with an inexact measurement of the state [19]. When dealing with a time series of data points $x_1, x_2, ..., x_n$ a filter computes the best guess for the point x_{n+1} taking into account all previous points and provides a correction using an inexact measurement of x_{n+1}.

The next section explains the architecture and structure of the proposed system. The experimental evaluation is presented in section 3. In this section we apply our system to estimate the airflow of a person with Sleep Apnoea. We use the on-line change detection algorithm to detect changes in the airflow. The last section presents the conclusions and lessons learned.

2   Detection System in Regression Models with Kalman Filter and CUSUM

In this paper we propose a modular detection system (DSKC) for regression problems. The general framework is shown in Figure 2.2. The system is composed by three components: a regression learning algorithm, a Kalman filter [15] and a CUSUM [2,4].

At each iteration, the system's first component, the learning algorithm, receives one unlabeled example, x_i, and then the actual model predicts, $\hat{y}_i$. After the model forecast, it receives an input from the environment y_i and calculates the residual $r_i = (y_i - \hat{y}_i)$. The system uses r_i and the Kalman filter

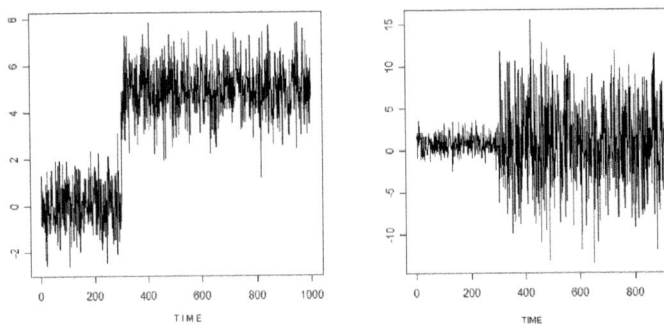

Fig. 1. Two types of concept drift: changes in the mean and changes in dispersion

error estimate of the actual model, $\hat{r}_i$, to compute a residual for the variance, $rv_i = (r_i - \hat{r}_i)^2$.

A summary of the process is:

1. the regression algorithm receives an unlabeled example, (x_i), and makes a prediction, $\hat{y}_i$;
2. after the prediction, the system receives the feedback from the environment y_i;
3. the system computes the residual for that example, $r_i = (y_i - \hat{y}_i)$;
4. the actual mean error, $\hat{r}_i$, is estimated by the kalman filter;
5. the system computes the residual of the variance for that example using the Kalman filter estimate for the mean error, $rv_i = (r_i - \hat{r}_i)^2$;
6. the actual variance error, $\hat{r}v_i$ is estimated by the Kalman filter;
7. both residuals are compared are compared with the estimates predict by the Kalman filter for the current state of the system by a *two side CUSUM's*;
   (a) a change is detected if there is a significant difference between the prediction of the state of the residuals and the current state predict by the Kalman filter. In that case the regression model is re-initialized with the new examples.
   (b) otherwise, the Kalman filters updates its own state.

2.1   Kalman Filter

The Kalman filter is a recursive filter that estimates the state of a dynamic system from a series of noisy measurements. The Kalman filter receives both residuals and updates the learning algorithm state estimate. The state estimate is formed by the mean error, $\hat{r}_i$, and the dispersion error, $\sigma_{r_i}^2$. Normally, a learning algorithm will improve the predictions with the arrival of new examples, mainly in the initial learning stage. For that reason, it is very important to provide a run-time estimation of the residuals. In general, run-time estimation is provided by a simple mechanism, such as auto regressive or auto regressive moving average or Kalman filter. The advantages of this last filter are: it allows to adaptively tune the filter memory to faster track variations in the estimation and allows to improve the accuracy of the estimation by exploiting the state update laws and variance of the estimation [15]. Assume that at time stamp $i-1$ the current error of the regression model is $\hat{r}_{i-1}$. When a new labeled example is available, the regression model predicts $\hat{y}_i$ and the system computes the residual $r_i = (\hat{y}_i - y_i)$. The Kalman filter combines both error measures to give a better estimate of the true error:

$$\hat{r}_i = w\hat{r}_{i-1} + (1 - w)r_i \tag{1}$$

so the variance of $\hat{r}_i$, $\sigma_{\hat{r}_i}^2$, is given by:

$$\sigma\hat{}_{r_i} = w^2\sigma_{\hat{r}_{i-1}}^2 + (1 - w)^2\sigma_{r_i}^2. \tag{2}$$

The weight that minimizes the variance of the new estimate of the true error, is

$$w = \frac{\sigma_{r_i}^2}{\sigma_{\hat{r}_{i-1}}^2 + \sigma_{r_i}^2}. \tag{3}$$

Hence, (1), (2) become:

$$\hat{r}_i = \frac{\frac{\hat{r}_{i-1}}{\sigma_{\hat{r}_{i-1}}^2} + \frac{r_i}{\sigma_{r_i}^2}}{\sigma_{\hat{r}_{i-1}}^2 + \sigma_{r_i}^2} \tag{4}$$

$$\sigma_{\hat{r}_i}^2 = \frac{\sigma_{r_i}^2 \sigma_{\hat{r}_{i-1}}^2}{\sigma_{\hat{r}_{i-1}}^2 + \sigma_{r_i}^2} \tag{5}$$

After some algebraic manipulation the equations 4 and 5 can be re-written as:

$$\hat{r}_i = \hat{r}_{i-1} + K(r_i - \hat{r}_{i-1}) \tag{6}$$

and

$$\sigma_{\hat{r}_i}^2 = (1 - K)\sigma_{\hat{r}_{i-1}}^2 \tag{7}$$

where

$$K = \frac{\sigma_{\hat{r}_{i-1}}^2}{\sigma_{\hat{r}_{i-1}}^2 + \sigma_{r_i}^2} \tag{8}$$

equations 6, 7, and 8 are the univariate Kalman filter. These are the equations are used to estimate the error mean value the system.

The error variance of the system, $\sigma_{r_i}^2$, can be estimate using the same process but in this case r_i is substitute by $rv_i = (rv_i - \hat{r}v_i)^2$ and $\hat{r}_{i-1}$ is $\hat{r}v_{i-1}$. The $\sigma_{rv_i}^2$ is estimated recursively by the system.

2.2   The Cumulative Sum Algorithm

The cumulative sum (CUSUM algorithm) is a sequential analysis technique typically used for monitoring change detection [16]. There are several versions of the CUSUM algorithm. The version used in this study is the Wald's CUSUM Sequential Probability Ratio Test (SPRT) algorithm [8]. The classical CUSUM detects changes in the mean value of a random variable. The detection algorithm gives an alarm when the cumulative sum of log-likelihood ratio (S) shows significant differences between the following hypotheses $H_0 : \mu = \mu_0$ vs $H_1 : \mu = \mu_1 = \mu_0 + \nu$, where μ is the mean and ν the magnitude of change allowed. The cumulative sum of log-likelihood ratio is the sum of the ratios log between the probability of both hypotheses,

$$S = \sum s_i \text{ where } s_i = log\left(\frac{p_{\mu_1}}{p_{\mu_0}}\right).$$

If μ_0 is true then p_{μ_0} is higher than p_{μ_1} and consequently s_i is lower than 0 on the other end, if μ_1 is true then p_{μ_0} is lower than p_{μ_1} and consequently s_i is higher than 0. Taking into account this result, it is possible to construct an adaptive window size strategy. If S_i is lower than ϵ, then S_i is restarted $(S_i = 0)$, and the window size is 0; if $\epsilon < S_i < h$ then the window size grows by one; If $S_i > h$ then an alarm is signaled by the CUSUM.

Assuming that the input has a normal distribution with mean μ and variance σ^2 then the density function is:

$$p_\mu\left(r\right) = \frac{1}{\sigma\sqrt{2\pi}} \exp^{-\frac{(y-\mu)^2}{2\sigma^2}}, \tag{9}$$

so in the previous hypothesis, s_i is equal to:

$$s_i = log\left(\frac{p_{\mu_1}\left(r\right)}{p_{\mu_0}\left(r\right)}\right) = \frac{\mu_1 - \mu_0}{\sigma^2}\left(y_i - \frac{\mu_1 + \mu_0}{2}\right)$$

The classical CUSUM can be used to detect changes in the variance σ^2: the null hypotheses is $\sigma = \sigma_0$ and the alternative hypotheses is $\sigma = \sigma_1$. In that case, s_i is equal to:

$$s_i = log\left(\frac{p_{\sigma_1}\left(r\right)}{p_{\sigma_0}\left(r\right)}\right) = ln\left(\frac{\sigma_0}{\sigma_1}\right) + \left(\frac{1}{\sigma_0^2} - \frac{1}{\sigma_1^2}\right)\frac{(y_k - \mu)^2}{2}$$

The input data for CUSUM can be any residual or any filter residual, for instance the prediction error from a Kalman filter. The CUSUM test is memoryless, and its accuracy depends on the choice of parameters v and h which control the tradeoff between false positive rate and true detection.

The proposed system detects changes in mean error of the actual model or a change in the respective variance. Figure 1 illustrates the two types of changes we are interested in. The pair $(r_i, \hat{r}_i)$ is transmitted to the mean CUSUM and the pair $(rv_i, \hat{r}v_i)$ is transmitted to the variance CUSUM. Both CUSUM's compare the values they receive. A change occurs if significant differences between both values received or significant differences between consecutive residuals are found. In both cases, a change in the mean or in the variance, the system gives an order to erase the actual learned model and start to construct a new model using the new examples. If no significant differences are found, the new example is incorporated into the learning model.

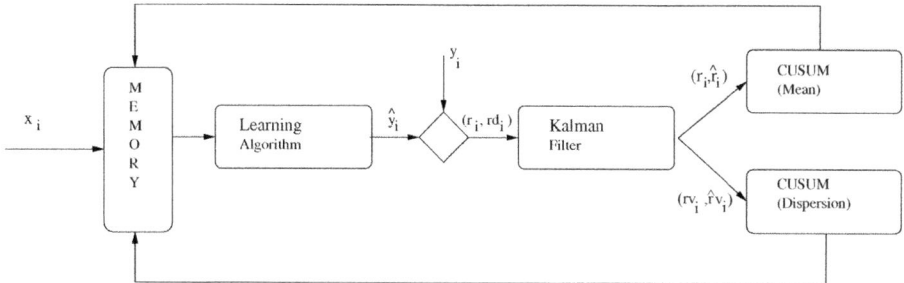

Fig. 2. Detection System with Kalman Filter and CUSUM Framework

The decision rule for CUSUM is:

$$\left\{ g_k = S_{k-N_k}^k \geq h \right\}$$

where $N_k = N_{k-1} \cdot 1_{g_{k-1} > \epsilon} + 1$. The algorithm signals an alarm, indicating a change, when $t_a = \min \{k : S_k \geq h\}$.

Initializing the parameters of the Algorithm. In order to complete the description of the algorithm, it is necessary to explain how to initialize the following parameters $\hat{r}_i$, $\sigma_{r_i}^2$, $\sigma_{\hat{r}_i}^2$, $\hat{r}v_i$, $\sigma_{rv_i}^2$, $\sigma_{\hat{r}v_i}^2$. We use the first 30 residuals from the model generated after a change is detected (and at the beginning) to estimate these parameters. The $\hat{r}_i$ and $\hat{r}v_i$ are the mean, the $\sigma_{r_i}^2$ and $\sigma_{rv_i}^2$ are the variance of both type of residuals (r and rv), respectively. Finally, $\sigma_{\hat{r}_i}^2$ and $\sigma_{\hat{r}v_i}^2$ are equal to the variance estimated for both residuals divided by 30. The minimal change considered in the mean was $\nu = \pm 6\sigma_0$ and for the variance was $\nu = 2$ or $\nu = 1/2$.

2.3   Discussion

The proposed architecture is general. It can work with any regression learning algorithm and any loss function. The main assumption is that a change is reflected in the distribution of the examples, leading to an increase of the error of the actual regression model.

3   Experimental Evaluation

In this section we describe the evaluation of the proposed system. We used three artificial datasets and two real-world datasets. A real-world physiological [1] dataset was used to evaluate if our DSKC provides a reliable estimate of the learning model error.

We used three artificial datasets and one real-world dataset of Sleep Apnoea to evaluate the efficiency of the DSKC. Artificial data allows us to perform controlled experiments. The experiments were designed in a way that we know when the change occurs.

To test the generality of the proposed methodology, we used two distinct learning algorithms: a regression tree and a linear regression model[1].

Four performance measures were used to evaluate the efficiency of the DSKC: number of false alarms (FA) and true alarms (TA), mean number of examples for detection of the changes (MNE) and normalized mean absolute error (NMAE). The median test [3] was used to compare the normalized mean absolute error with and without change detection for each dataset and learning algorithm.

3.1   Physiological Dataset

In this section we study the impact in boosting the discriminative power of a signal given by our system components: the Kalman filter and the CUSUM. We use the physiological dataset [1] that was collected using BodyMedia wearable body monitors. These continuous data are measurements of 9 sensors and an indication of the physical activities of the user. The dataset comprises several months of data from 18 subjects.

We divided the dataset by subject and age and used only the sets related to changes between sleep and awake. We measured the discrimination between sleep and awake phases using the original sensors, using only the Kalman filter estimate of a sensor, and using the estimate of the sensor from Kalman filter with CUSUM. We used the sensors 7 and 9 because they were the sensors with the larger and the smaller discriminative power, respectively. The discrimination power was measured using the area under the ROC curve.

The results show (Table 1) that the discrimination power increases when we applied the Kalman filter or the Kalman filter with CUSUM to the original sensors. This fact is more evident when the sensor exhibit less discriminative power. The less discriminative sensor is sensor 9. In that case, the improvement verified with the Kalman filter plus the CUSUM is 5.9% with a p-value $p < 0.002$. The improvement decreases for sensor 7, where the Kalman filter alone has better results ($2.2\%, p < 0,018$). These results suggest that the use of the Kalman filter with CUSUM provides a reliable estimate of the learning model error.

---

[1] In the set of experiments reported here, we use batch versions of both algorithms that train a new model at every iteration using all examples from the last change detected or since the beginning until that moment. The focus of the paper is change detection in regression problems. It is expected that the main conclusions apply to incremental versions of the algorithms.

Table 1. Area under the ROC curve for the sensor, the Kalman filter estimate (KF) and Kalman filter with CUSUM estimate (KFC)

Area Under the ROC Curve						
	sensor 7			sensor 9		
ID	KFC	KF	sensor	KFC	KF	sensor
1	0.88	0.93	0.92	0.78	0.81	0.73
2	0.95	0.95	0.87	0.81	0.76	0.66
3	0.96	0.96	0.89	0.83	0.77	0.67
4	0.95	0.97	0.92	0.65	0.60	0.58
5	0.96	0.96	0.91	0.72	0.66	0.59
6	0.92	0.94	0.91	0.85	0.82	0.75
7	0.92	0.92	0.89	0.85	0.81	0.73
8	0.91	0.89	0.87	0.60	0.57	0.50
9	0.88	0.89	0.86	0.68	0.65	0.56
10	0.81	0.93	0.85	0.59	0.58	0.55
11	0.94	0.99	0.99	0.62	0.63	0.57
12	0.95	0.97	0.97	0.68	0.64	0.56
13	0.94	0.97	0.93	0.85	0.84	0.83
14	0.93	0.97	0.97	0.68	0.58	0.53
15	0.92	0.92	0.85	0.62	0.59	0.53
16	0.95	0.97	0.94	0.53	0.50	0.547
Median	**0.94**	**0.95**	**0.91**	**0.68**	**0.65**	**0.57**

3.2 Artificial Datasets

The three artificial datasets used were composed by 3000 random examples. Two random changes were generated between the 30th and the 2700th examples.

1. The first dataset has five normally distributed attributes with mean 0 and standard deviation 50. The dependent variable is a linear combination of the five attributes with white noise with standard deviation 10. New coefficients of the linear combination where built at every change. The five coefficients of the linear combination where generated by a uniform distribution over [0, 10].
2. The second dataset has two uniformly distributed attributes over [0, 2]. The dependent variable is a linear combination of the first sine attribute, and the cosine of the second attribute. We add, to each attribute, white noise with standard deviation 1. As in the previous artificial dataset, new coefficients were built at every change.
3. The third dataset is a modified version of the dataset that appear in the MARS paper [9]. This dataset has 10 independent predictor variables $x_1, ..., x_{10}$ each of which is uniformly distributed over [0,1]. The response is given by

$$y = 10 \sin(\pi x_1 x_2) + 20(x_3 - 0,5)^2 + 10x_4 + 5x_5 + \epsilon \cdot \quad (10)$$

A permutation of the predictor variables was made at each change.

For each type of dataset we randomly generated 10 datasets.

3.3   Results on Artificial Datasets

Tables 2, 3 and 4 show the results for dataset one, two, and three, respectively. We can observe that DSKC is effective with all learning algorithms. Overall we detected 73% ($CI95\% = [64\% - 81\%]$) of all true changes with no false alarms ($CI95\% = [0\% - 6\%]$). The results show that the proportion of true changes varies between 50% for the third dataset with linear regression and 90% for the same dataset but with regression trees; the mean number examples needed for detection varies from 8.3 to 42.13.

Table 2. Results for the first artificial dataset

	Regression Trees				Linear Models				
	No Detection	Detection			No Detection	Detection			
	NMAE	NMAE	TA	FA	MNE	NMAE	NMAE	TA	FA	MNE
1	0.75	0.71	1	0	45.0	0.30	0.10	2	0	3.0
2	0.73	0.62	2	0	38.0	0.25	0.11	1	0	9.0
3	0.85	0.66	2	0	27.5	0.51	0.11	2	0	5.0
4	0.68	0.67	1	0	16.0	0.29	0.12	1	0	4.0
5	0.66	0.63	2	0	45.0	0.40	0.13	2	0	19.5
6	0.68	0.64	2	0	40.5	0.31	0.10	2	0	2.0
7	0.79	0.57	2	0	9.5	0.30	0.21	1	0	6.0
8	0.73	0.59	1	0	51.0	0.28	0.08	2	0	26.5
9	0.73	0.69	1	0	43.0	0.22	0.08	2	0	4.5
10	0.84	0.76	1	0	10.0	0.38	0.09	2	0	3.5
η	0.73	0.65	2	0	39.2	0.30	0.11	2	0	4.8
$\bar{x}$	0.74	0.65	1.5	0	32.6	0.32	0.11	1.7	0	8.3

We found significant differences ($p < 0.05$) between the use and not use of our detection system for the normalized mean absolute error, except for the 2nd dataset with the linear regression model. The mean normalized error decreased for all datasets with the use of our DSKC. We observed that when the second change occurs relatively close to the first change or when the first change occurs relatively close to the beginning of the experience, the change was not detected. As we can see in Table 5, the proportion of changes detected was 25%, when the number of examples between changes is less than 332, against 89%, when there

Table 3. Results for the second artificial dataset

	Regression trees					Linear models				
	No Detection	Detection				No Detection	Detection			
	NMAE	NMAE	TA	FA	MNE	NMAE	NMAE	TA	FA	MNE
1	0.57	0.56	1	0	8.0	0.85	0.82	1	0	23.0
2	0.45	0.33	1	0	5.0	0.88	0.88	2	0	48.0
3	0.45	0.42	1	0	78.0	0.89	0.89	1	0	35.0
4	0.53	0.47	2	0	10.5	0.81	0.82	2	0	95.0
5	0.39	0.37	1	0	13.0	0.90	0.90	2	0	84.5
6	0.61	0.36	2	0	9.0	0.83	0.83	0	0	–
7	0.48	0.39	1	0	39.0	1.00	1.00	1	0	5.0
8	0.54	0.45	2	0	8.0	1.00	1.00	2	0	89.5
9	0.45	0.41	2	0	10.0	0.84	0.85	1	0	27.0
10	0.45	0.41	2	1	33.5	0.87	0.87	1	0	83.0
η	0.47	0.41	1.5	0	10.2	0.87	0.87	1.0	0	48.0
$\bar{x}$	0.49	0.42	1.5	0	21.4	0.89	0.88	1.3	0	54.4

Table 4. Results for the third artificial dataset

	Regression trees					Linear models				
	No Detection	Detection				No Detection	Detection			
	NMAE	NMAE	VA	FA	MNE	NMAE	NMAE	VA	FA	MNE
1	0.80	0.67	2	0	21.5	0.71	0.59	1	0	38.0
2	0.73	0.69	1	0	33.0	0.73	0.73	0	0	–
3	0.84	0.66	2	0	23.0	0.68	0.57	1	0	65.0
4	0.86	0.67	2	0	28.0	0.71	0.58	2	0	55.0
5	0.82	0.66	2	0	33.0	0.68	0.57	1	0	19.0
6	0.71	0.68	1	0	14.0	0.75	0.59	2	0	54.5
7	0.86	0.68	2	0	39.0	0.69	0.59	1	0	25.0
8	0.80	0.66	2	0	50.5	0.63	0.60	1	0	41.0
9	0.87	0.68	2	0	20.5	0.88	0.88	0	0	–
10	0.82	0.68	2	0	17.5	0.67	0.59	1	0	39.0
η	0.82	0.67	2	0	25.5	0.70	0.59	1.0	0	40.0
$\bar{x}$	0.81	0.67	1.8	0	28.0	0.63	0.62	1.0	0	42.1

Table 5. Association between the number of examples read and the ability to detect versus not detect changes

Number Examples	Not Detected	Detected	p
$[1 - 332[$	9 (75.0%)	3 (25.0%)	< 0.001
$[332 - 532[$	11 (45.8%)	13 (54.2%)	
$[532 - \infty[$	8 (11.1%)	64 (88.9%)	

are more than 532 examples. The association between the number of examples required by the learning algorithm and detection or not detection of the change is significant ($p < 0.001$).

3.4   Sleep Apnoea Dataset

After measuring the performance of our detection system in the artificial datasets, we evaluated the performance on a real problem where change points and change rates are not known. For such we applied our system to a dataset from patients with sleep apnoea. Sleep apnoea is a common disorder characterized by periods of breathing cessation (apnoea) and periods of reduced breathing (hyponea). The standard approach to diagnoses apnoea consists of monitoring a wide range of signals (airflow, snoring, oxygen saturation, heart rate...) during patient sleep. There are several methods for quantifying the severity of the disorder, such as measuring the number of apnoeas and hypopnoea per hour of sleep or measuring the number of breathing events per hour. There is heterogeneity of methods for defining abnormal breathing events, such as reduction in airflow or oxygen saturation or snoring [7]. It can be seen as pathological, when the number of apnoeas and Hypopnoea/hour is larger than 20 events per hour [12]. Our goal in this experiment was to evaluate if our detection system could detect abnormal breathing events.

The real-world dataset was a set of sleep signals from a patient with sleep apnoea. Three of the 7 signals (airflow, abdominal movement signals and snoring) had 16Hz frequency and the other 4 signals (heart rate, light, oxygen saturation and body) had 1Hz frequency. All signals with 16Hz were transform in 1Hz using the mean for each second. The dataset contained 26102 records from 7 signals, which is approximated 7.5 hours of sleep.

3.5   Results on Sleep Apnoea Dataset

Taking into consideration the problem, we built a model to predict the airflow using all other signals as predictor variables. In this dataset, the regression model is evaluated using the normalized mean absolute error statistic. We did not have any indication where the abnormal breathing event would occur. For this reason, we cannot evaluate the number of true and false alarms neither the mean number of examples for detection of the changes.

We used two regression models: a generalized linear regression and a regression tree[2]. Both learning algorithms employed exhibited slight better results using the detection mechanism than without (Table 6). The total number of alarms (TA) was 18 and 17, for the regression tree and linear regression models, respectively.

In order to evaluate the agreement between both learning algorithms to detected abnormal breathing events, we compared the alarms proximity between them. We detected 13 pairs of alarms, specifically, 13 alarms detected by the regression tree model occurred in the proximity of 13 alarms detected by the linear

---

[2] The GLM and CART versions implemented in [18].

regression model (Table 7). To validate how well the CUSUM detect changes, we carried out a second set of experiments. We design two datasets. In both datasets, we consider only a single attribute: the time-Id. The target variable, in problem 1, is the mean of the airflow predicted by the Kalman Filter. In problem 2, the target variable is the dispersion of the airflow predict by the Kalman Filter. We run a regression tree in both problems and collect the cut-points from both trees. The set of cut-points of the regression trees and the alarms detected by our DSKC were compared (Figure 3).

Table 6. Results for dataset of Sleep Apnoea

	Regression Trees			Linear Model		
	No Detection	Detection		No Detection	Detection	
	NMAE	NMAE	TA	NMAE	NMAE	TA
1	0.940	0.923	18	0.981	0.974	17

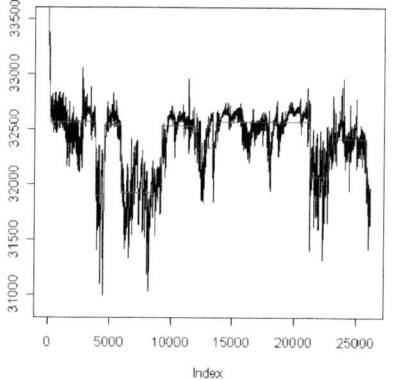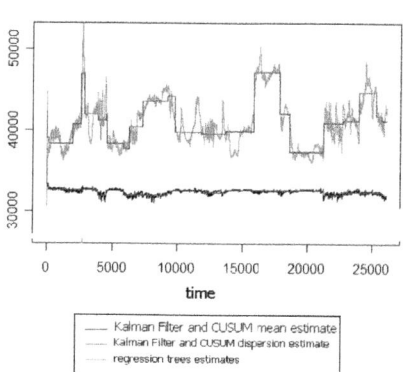

Fig. 3. The regression tree segmentation of the mean airflow (left) and airflow dispersion (right) filter by Kalman Filter and CUSUM

As shown in Table 7, there were 7 increases detected in time in the airflow dispersion and all of them were detected at least by one of the learning models applied to sleep apnoea dataset. There were 4 increases detected in time in the airflow, and 2 of them by one of the learning models applied. Despite both learning algorithms investigated exhibited slightly better results using the detection mechanism than without, the alarms detected by both models seem to show agreement, which may imply that we have detected true changes in the distribution of examples in the sleep apnoea dataset.

Table 7. Alarms detected by DSKC with regression trees and linear models in sleep apnoea dataset and the cut points of a regression tree trained off-line

	Regression trees	Linear model	Cut-points (Reg. tree)	type of change
1	2017	1965	1962.5	disp,+
2	2600	2682	2631.5	disp,+
3	—	—	2851.5	disp,-
4			3864.5	mean,-
5	4172	4135	—	—
6	—	—	4551.5	mean,+
7	—	—	4595.5	disp,-
8	5835	—	—	—
9	—	—	5875.5	mean,-
10	—	6165	6324.5	disp,+
11	—	7415	7322.5	disp,+
12	—	9287	9202.5	mean,+
13	—	—	9764.5	disp,-
14	10207	10211	—	—
15	11106	11112	—	—
16	11531	—	—	—
17	—	—	11793.5	mean,-
18	12318	12452	—	—
19	13404	13396	13632.5	mean,+
20	14686	14927	—	—
21	15848	15802	15808.5	disp,+
22	—	17762	—	—
23	—	—	17833.5	disp,-
24	18046	—	—	—
25	—	—	18609.5	disp,-
26	20456	20463	—	—
27	—	—	21207.5	mean,-
28	21216	21280	21222.5	disp,+
29	22505	22253	—	—
30	—	—	22743.5	mean,+
31	23139	—	—	—
32	24018	—	23961.5	disp,+
33	24581	24400	—	—
34	—	—	25290.5	disp,-
35	—	—	25733.5	mean,-

4   Conclusions

In this paper we discussed the problem of maintaining accurate regression models in dynamic, non-stationary environments. The system continuously monitors the residual of the regression algorithm, looking for changes in the mean and changes in the variance. The proposed method maintains a regression model where residuals are filtered by a Kalman filter. A CUSUM algorithm continuously monitors significant changes in the output of the Kalman filter. The CUSUM works as a wrapper over the learning algorithm (the regression model plus the Kalman filter), monitoring the residuals of the actual regression model. If CUSUM detects an increase of the error, a new regression model is learned using only the most recent examples. As shown in the experimental section the Kalman filter application to the residuals gives a good on-line estimation of the learning algorithm state. The results of the method for change detection in regression problems show that it can be applied with efficiency when the information is available sequentially over time. An advantage of the proposed method is that it is independent of the learning algorithm. The results of the change detection algorithm mainly depend on the efficiency of the learning algorithm. They also show that the Kalman filter has a good performance in detecting real changes from noisy data.

Acknowledgments. Thanks to Andre Carvalho for useful comments and to the financial support given by the FEDER, the Plurianual support attributed to LIAAD, and project Knowledge Discovery from Ubiquitous Data Streams (PTDC/EIA-EIA/098355/ 2008).

References

1. Andre, D., Stone, P.: Physiological data modeling contest. Technical report, University of Texas at Austin (2004)
2. Basseville, M., Nikiforov, I.: Detection of Abrupt Changes: Theory and Applications. Prentice-Hall, Englewood Cliffs (1993)
3. Bhattacharyya, G., Johnson, R.: Statistical Concepts and Methods. John Willey & Sons, New York (1977)
4. Bianchi, G., Tinnirello, I.: Kalman filter estimation of the number of competing terminals in IEEE. In: The 22nd Annual Joint Conference of IEEE Computer and Communications (2003)
5. Cauwenberghs, G., Poggio, T.: Incremental and decremental support vector machine learning. Advances in Neural Information Processing Systems 13 (2001)
6. Domingos, P., Hulten, G.: Mining high-speed data streams. In: Knowledge Discovery and Data Mining, pp. 71–80 (2000)
7. Flemons, W.W., Littner, M.R., Rowley, J.A., Anderson, W.M., Gay, P., Hudgel, D.W., McEvoy, R.D., Loube, D.I.: Home diagnosis of sleep apnoeas: A systematic review of the literature. Chest, 1543-1579, 211–237 (2003)
8. Wald, A.: Sequential Analysis. John Wiley and Sons, Inc., Chichester (1947)
9. Friedman, J.: Multivariate adaptive regression splines. Annals of Statistics 19(1), 1–141 (1991)

10. Gama, J., Medas, P., Castillo, G.: Learning with drift detection. In: Bazzan, A.L.C., Labidi, S. (eds.) SBIA 2004. LNCS (LNAI), vol. 3171, pp. 286–295. Springer, Heidelberg (2004)
11. Grant, E., Leavenworth, R.: Statistical Quality Control. McGraw-Hill, New York (1996)
12. Guimares, G., Peter, J.H., Penzel, T., Ultsch, A.: A method for automated temporal knowledge acquisition applied to sleep-related breathing disorders. Artificial Intelligence in Medicine 23, 211–237 (2001)
13. Higgins, C.M., Goodman, R.M.: Incremental learning using rule-based neural networks. In: International Joint Conference on Neural Networks, Seattle, WA, pp. 875–880 (1991)
14. Hulten, G., Spencer, L., Domingos, P.: Mining time-changing data streams. In: Proceedings of Knowledge Discovery and Data Mining. ACM Press, New York (2001)
15. Kalman, R.E.: A new approach to linear filtering and prediction problems. Transaction of ASME - Journal of Basic Engineering, 35–45 (1960)
16. Page, E.S.: Continuous inspection schemes. Biometrika 41(1/2), 100–115 (1954)
17. Pang, K.P., Ting, K.M.: Improving the centered CUSUMs statistic for structural break detection in time series. In: Proc. 17th Australian Joint Conference on Artificial Intelligence. LNCS. Springer, Heidelberg (2004)
18. R Development Core Team. R: A language and environment for statistical computing. R Foundation for Statistical Computing, Vienna, Austria (2005), ISBN: 3-900051-07-0
19. Rojas, R.: The Kalman filter. Technical report, Freie University of Berlin (2003)
20. Welch, G., Bishop. G.: An introduction to the Kalman filter. Technical report, 95-041, Department of Computer Science, Department of Computer Science, University of North Caroline, Chapel Hill (April 2004)

A Geometric Approach to Monitoring Threshold Functions over Distributed Data Streams

Izchak Sharfman[1], Assaf Schuster[1], and Daniel Keren[2]

[1] Faculty of Computer Science,
Technion – Israel Institute of Technology,
Haifa, Israel
[2] Department of Computer Science,
Haifa University,
Haifa, Israel
tsachis@cs.technion.ac.il, assaf@cs.technion.ac.il, dkeren@cs.haifa.ac.il

Abstract. Monitoring data streams in a distributed system is the focus of much research in recent years. Most of the proposed schemes, however, deal with monitoring simple aggregated values, such as the frequency of appearance of items in the streams. More involved challenges, such as the important task of feature selection (e.g., by monitoring the information gain of various features), still require very high communication overhead using naive, centralized algorithms.

We present a novel geometric approach by which an arbitrary global monitoring task can be split into a set of constraints applied locally on each of the streams. The constraints are used to locally filter out data increments that do not affect the monitoring outcome, thus avoiding unnecessary communication. As a result, our approach enables monitoring of arbitrary threshold functions over distributed data streams in an efficient manner.

We present experimental results on real-world data which demonstrate that our algorithms are highly scalable, and considerably reduce communication load in comparison to centralized algorithms.

1 Introduction

A common requirement in many emerging applications is the ability to process, in real time, a continuous high volume stream of data. Examples of such applications are sensor networks [2], real-time analysis of financial data [1,18], and intrusion detection. These applications are commonly referred to as *data stream systems* [4]. The real-time nature of data stream systems and the vast amounts of data they are required to process introduce new fundamental problems that are not addressed by traditional Database Management Systems (DBMS). Traditional DBMS are based on a *pull* paradigm, where users issue queries regarding data stored by the system, and the system processes these queries as they are issued and returns results. Data stream systems [7,8,13,2,9] are based on a *push* paradigm, where the users issue *continuous queries* [5,17] that specify

M. May and L. Saitta (Eds.): Ubiquitous Knowledge Discovery, LNAI 6202, pp. 163–186, 2010.

the required processing of the data, which the system processes as it arrives, continuously providing the user with updated results.

Various types of continuous queries have been studied in the past, including continuous versions of selection and join queries [14], various types of aggregation queries [16,3], and monitoring queries [7]. While most previous work regarding data stream systems considers sequential setups (the data is processed by a single processor), many data stream applications are inherently distributed: examples include sensor networks [2], network monitoring [11], and distributed intrusion detection.

A useful class of queries in the context of distributed data streams are monitoring queries. Previous work in the context of monitoring distributed data streams considered monitoring simple aggregates, such as detecting when the sum of a distributed set of variables exceeds a predetermined threshold [11], or finding frequently occurring items in a set of distributed streams [15]. Some work has been done on monitoring more complex constructs derived from distributed streams, but the proposed solutions are customized for the problem at hand. Examples include [6] which presents an algorithm for detecting similar sets of streams among a large set of distributed streams, and [10], which presents an algorithm for approximating quantiles over distributed streams.

A useful, more general type of monitoring query can be defined as follows: let $X_1, X_2,...,X_d$ be frequency counts for d items over a set of streams. Let $f(X_1,X_2,...,X_d)$ be an arbitrary function over the frequency counts. We are interested in detecting when the value of $f(X_1,X_2,...,X_d)$ rises above, or falls below, a predetermined threshold value. We refer to this query as a threshold function query.

There is a fundamental difference between the cases of linear and non-linear f, which can be demonstrated even for the case of one-dimensional data. Let x_1 and x_2 be values stored in two distinct nodes, and let $f(x) = 6x - x^2$. Suppose one needs to determine whether $f(\frac{x_1+x_2}{2}) > 1$. If f was linear, the solution would have been simple, since in that case $f(\frac{x_1+x_2}{2}) = \frac{f(x_1)+f(x_2)}{2}$. Suppose that initially the value at each node is < 1; then a simple distributed algorithm for monitoring whether $f(\frac{x_1+x_2}{2}) > 1$ is for each node i to remain silent as long as $f(x_i) < 1$. However, even for the simple non-linear function above, *it is impossible to determine from the values of f at the nodes whether its value at the average is above 1 or not.* For example, if $x_1 = 0, x_2 = 6$, then f's value in each node is below 1, but its value in the average of x_1 and x_2 is 9. But if $x_1 = 10, x_2 = 20$, the value at both x_i and their average is below 1. So, nothing can be deduced about the location of $f(\frac{x_1+x_2}{2})$ vis-a-vis the threshold given the locations of $f(x_i)$ vis-a-vis it.

In this trivial example, the cost of sending the data stored in the nodes is the same as sending the value, but in data mining applications the data can be of very high dimensionality. This necessitates a distributed algorithm for locally determining whether f's value at the average data vector is above the threshold.

Following is a more practical example of a threshold function query: consider a classifier built over data extracted from a set of streams, for example a distributed spam mail filtering system. Such a system is comprised of agents installed on several dispersed mail servers. Users mark spam mail they have received as such, providing each server with a continuous stream of positive and negative samples. These samples serve as a basis for building a classifier. Since the vocabulary comprising these samples may be very large, an important task in such a setup is determining which words, or features, should be used for performing the classification. This task is known as feature selection. Feature selection is typically performed by calculating, for every feature, a non-linear scoring function, such as Information Gain or χ^2, over statistics collected from all the streams. All the features scoring above a certain threshold are chosen as parameters for the classification task. Since the characteristics of spam mail may vary over time, one may wish to monitor the features in order to determine if selected features remain prominent, or if any of the features not selected have become prominent. Determining whether a certain feature should be selected at a given time can be viewed as a threshold function query.

Threshold function queries can be implemented by collecting all the mail items to a central location, but such a solution is very costly in terms of communications load. We are interested in algorithms that implement threshold function queries in a more efficient manner. We achieve this by defining numerical constraints on the data collected at each node. As data arrives on the streams, every node verifies that the constraint on its stream has not been violated. We will show that as long as none of these constraints have been violated, the query result is guaranteed to remain unchanged, and thus no communication is required.

Here we present two algorithms for efficiently performing threshold function queries. The algorithms are based on a geometric analysis of the problem. Upon initialization, the algorithms collect frequency counts from all the streams, and calculate the initial result of the query. In addition, a numerical constraint on the data received on each individual stream is defined. As data arrives on the streams, each node verifies that the constraint on its stream has not been violated. The geometric analysis of the problem guarantees that as long as the constraints on all the streams are upheld, the result of the query remains unchanged, and thus no communication is required. If a constraint on one of the streams is violated, new data is gathered from the streams, the query is reevaluated, and new constraints are set on the streams.

The first algorithm is a decentralized algorithm, designed for a closely coupled environment, where nodes can efficiently broadcast messages. The second algorithm is designed for loosely coupled environments, where the cost of broadcasting a message is high. These algorithms are, to the best of our knowledge, the first to enable efficient monitoring of arbitrary threshold functions over distributed data streams.

1.1   Detailed Example

Following is a detailed description of the spam filtering example given above. We will use this example to demonstrate the concepts we present. Let $p_1, p_2, ..., p_n$

be n agents installed on n different mail servers. Let $M_i = \{m_{i,1}, m_{i,2}, \dots, m_{i,k}\}$ be the last k mail items received at the mail server installed on p_i, and let M denote the union of the last k mail items received at each one of the n mail servers, $M = \bigcup_{i=1}^{n} M_i$. Let X denote a set of mail items, let $Spam(X)$ be the set of mail items in X labeled as spam, and let $\overline{Spam}(X)$ be the set of mail items in X not labeled as spam. Let $Cont(X, f)$ be the set of mail items in X that contain the feature f, and let $\overline{Cont}(X, f)$ be the set of mail items in X that do not contain the feature f. Let the contingency table $C_{f,X}$ for the feature f over the set of mail items X be a 2×2 matrix, $C_{f,X} = \{c_{i,j}\}$, such that $c_{1,1} = \frac{|Cont(X,f) \cap Spam(X)|}{|X|}$, $c_{1,2} = \frac{|Cont(X,f) \cap \overline{Spam}(X)|}{|X|}$, $c_{2,1} = \frac{|\overline{Cont}(X,f) \cap Spam(X)|}{|X|}$, and $c_{2,2} = \frac{|\overline{Cont}(X,f) \cap \overline{Spam}(X)|}{|X|}$. C_{f,M_i} is called the local contingency table for the node p_i, and $C_{f,M}$ is called the global contingency table. Note that $C_{f,M} = \frac{\sum_{i=1}^{n} C_{f,M_i}}{n}$. We are interested in determining, for each feature f, whether the Information Gain over its global contingency table, denoted by $G(C_{f,M})$, is above or below a predetermined threshold r. The formula for Information Gain is given below[1]

$$G(C_{f,X}) = \sum_{i \in \{1,2\}} \sum_{j \in \{1,2\}} c_{i,j} \cdot \log\left(\frac{c_{i,j}}{(c_{i,1} + c_{i,2}) \cdot (c_{1,j} + c_{2,j})}\right)$$

Note that the answer to the threshold function query *cannot* be derived from the value of the monitored function on data from each individual stream. Consider, for example, a spam filtering system consisting of two streams, with a threshold value of 0.5. The first node may hold a contingency table $C_{f,M_1} = \begin{pmatrix} 1 & 0 \\ 0 & 0 \end{pmatrix}$, resulting in $G(C_{f,M_1}) = 0$, and the second node may hold a contingency table $C_{f,M_2} = \begin{pmatrix} 0 & 0 \\ 0 & 1 \end{pmatrix}$, resulting in $G(C_{f,M_2}) = 0$. As we can see, the gain calculated on each individual stream is 0, and thus below the threshold value, but the gain on the global contingency table for f, $C_{f,M_1 \cup M_2} = \frac{C_{f,M_1} + C_{f,M_2}}{2} = \begin{pmatrix} 0.5 & 0 \\ 0 & 0.5 \end{pmatrix}$, is $G(C_{f,M_1 \cup M_2}) = 1$, and thus above the threshold value. Note that this behaviour does not occur when monitoring frequencies of occurrence of items over distributed streams, i.e., if the frequency of occurrence of a certain item in all the streams is below a predetermined threshold, then the frequency of occurrence of that item over the union of the streams is below the threshold as well.

2   Computational Model

Let $S = \{s_1, s_2, \dots s_n\}$, be a set of n data streams, collected at nodes $P = \{p_1, p_2, \dots, p_n\}$. Let $v_1(t), v_2(t), \dots, v_n(t)$ be d-dimensional real vectors derived

---

[1] If $c_{i,j} = 0$ then $c_{i,j} \cdot \log\left(\frac{c_{i,j}}{(c_{i,1} + c_{i,2}) \cdot (c_{1,j} + c_{2,j})}\right)$ is defined as 0.

from the streams (the value of these vectors varies over time). These vectors are called *local statistics vectors*. Let $w_1, w_2, ..., w_n$ be positive weights assigned to the streams.

The weight w_i assigned to the node p_i usually corresponds to the number of data items its local statistics vector is derived from. Assume, for example, that we would like to determine whether the frequency of occurrence of a certain data item in a set of streams is above a certain threshold value. In this case, the weight we assign to each node at time t is the number of data items received on the stream at time t (and $v_i(t)$ is a scalar holding the frequency of occurrence of the item in the stream s_i). In this setup weights change over time. A variant of the problem stated above is for each node to maintain the frequency of occurrence of the item in the recent N_i data items received on the stream (this is known as working with a sliding window of size N_i). In that case, the weight assigned to each node is the size of its sliding window. In this setup weights do not change over time. For the sake of clarity, we assume at first that the weights are fixed, and that they are known to all nodes. Later, we modify our algorithms to handle weights that vary with time.

Let $v(t) = \dfrac{\sum\limits_{i=1}^{n} w_i v_i(t)}{\sum\limits_{i=1}^{n} w_i}$. $v(t)$ is called the *global statistics vector*. Let $f : \mathbb{R}^d \to \mathbb{R}$ be an arbitrary function from the space of d-dimensional vectors to the reals. f is called the monitored function. We are interested in determining at any given time, t, whether or not $f(v(t)) > r$, where r is a predetermined threshold value.

We present algorithms for two settings, a decentralized setting and a coordinator-based setting. Algorithms in both settings construct a vector called the estimate vector, denoted by $e(t)$. The estimate vector is constructed from the local statistics vectors collected from the nodes at certain times, as dictated by the algorithms. The last statistics vector collected from the node p_i is denoted by v'_i. Each node remembers the last statistics vector collected from it. The estimate vector is the weighted average of the latest statistics vectors collected from the nodes, i.e., $e(t) = \dfrac{\sum\limits_{i=1}^{n} w_i v'_i}{\sum\limits_{i=1}^{n} w_i}$. From time to time, as dictated by the algorithm, an updated statistics vector is collected from one or more nodes, and the estimate vector is updated. At any given time the estimate vector is known to all nodes.

In the decentralized setting, when the algorithm dictates that a statistics vector should be collected from a node, the node broadcasts the statistics vector to the rest of the nodes. Each node keeps track of the last statistics vector broadcast by every node, and locally calculates the estimate vector. In the coordinator-based setting, we designate a coordinator node and denote it by p_1. The coordinator is responsible for collecting local statistics vectors from the nodes, calculating the estimate vector, and distributing it to the nodes.

In both settings, each node p_i maintains a parameter called the statistics delta vector. This vector is denoted by $\Delta v_i(t)$. The statistics delta vector held by the node p_i is the difference between the current local statistics vector and the last statistics vector collected from the node, i.e., $\Delta v_i(t) = v_i(t) - v'_i$.

In both settings, each node p_i also maintains a parameter called the *drift vector*. This vector is denoted by $u_i(t)$. The drift vector is calculated differently in each setting. In the decentralized setting the drift vector is a displacement of $\Delta v_i(t)$ in relation to the estimate vector,

$$u_i(t) = e(t) + \Delta v_i(t) \tag{1}$$

The coordinator-based algorithm employs a mechanism for balancing the local statistics vectors of a subset of the nodes. Consider the case where at a certain time t the statistics delta vector in two equally weighted nodes, p_i and p_j, cancel each other out: that is $\Delta v_i(t) = -\Delta v_j(t)$. We will see that balancing the local statistics vectors held by p_i and p_j can improve the efficiency of the algorithm. The coordinator facilitates this balancing by sending each node a *slack vector*, denoted by δ_i. The sum of the slack vectors sent to the nodes is 0. The drift vector held by each node is calculated as follows:

$$u_i(t) = e(t) + \Delta v_i(t) + \frac{\delta_i}{w_i} \tag{2}$$

In the decentralized algorithm, nodes communicate by broadcasting messages. The cost of performing a broadcast varies according to the networking infrastructure at hand. In the worst case broadcasting a message to n nodes requires sending n point to point messages. While the decentralized algorithm remains highly efficient even in those settings, in practice, the cost of broadcasting a message is significantly lower. Some networking infrastructures, such as wireless networks and Ethernet based networks, support broadcasting at the cost of sending a single message. In other cases efficient broadcasting schemes have been developed that significantly reduce the cost of broadcasting.

We assume that communication links are reliable, i.e., no messages are lost (otherwise standard methods for implementing reliability can be employed).

3   Geometric Interpretation

At the heart of the approach is the ability to decompose the monitoring task into local constraints on streams. As data arrives on the streams, each node verifies that the local constraint on its stream has not been violated. We will show that as long as none of these constraints have been violated, the query result is guaranteed to remain unchanged, and thus no communication is required. As demonstrated in Section 1, this cannot be done solely by observing the value of the monitored function on each stream. Therefore, an estimated global statistics vector, called the estimate vector, is known to all nodes. The estimate vector is said to be correct at a given time if the value of the monitored function on the estimate vector and the value of the monitored function on the global statistics vector at that time (this value is unknown to any singe node) are on the same side of the threshold. Given an initially correct estimate vector, our goal is to set local constraints on each stream such that as long as no constraints have been violated, the estimate vector remains correct, and thus no communication is required. The method for decomposing the monitoring task is based on the following, easily verifiable observation—at any given time the weighted average of the drift vectors held by the nodes is equal to the global statistics vector,

$$\frac{\sum\limits_{i=1}^{n} w_i u_i(t)}{\sum\limits_{i=1}^{n} w_i} = v(t) \tag{3}$$

We refer to Property (3) as the convexity property of the drift vectors. The geometric interpretation of Property (3) is that the global statistics vector is in the convex hull of the drift vectors held by the nodes,

$$v(t) \in \mathrm{Conv}(u_1(t), u_2(t), ..., u_n(t)) \tag{4}$$

This observation enables us to take advantage of Theorem 1 in order to decompose the monitoring task.

Theorem 1. Let $x, y_1, y_2, ..., y_n \in \mathbb{R}^d$ be a set of vectors in $\mathbb{R}^d$. Let $\mathrm{Conv}(x, y_1, y_2, ..., y_n)$ be the convex hull of $x, y_1, y_2, ..., y_n$. Let $B(x, y_i)$ be a ball centered at $\frac{x+y_i}{2}$ and with a radius of $\left\| \frac{x-y_i}{2} \right\|_2$ i.e., $B(x, y_i) = \{ z \, | \, \|z - \frac{x+y_i}{2}\|_2 \le \left\| \frac{x-y_i}{2} \right\|_2 \}$, then $\mathrm{Conv}(x, y_1, y_2, ..., y_n) \subset \bigcup\limits_{i=1}^{n} B(x, y_i)$.

Theorem 1 is used to bound the convex hull of $n+1$ vectors in $\mathbb{R}^d$ by the union of n d-dimensional balls. In our case it is used to bound the convex hull of the estimate vector and the drift vectors i.e., $\mathrm{Conv}(e(t), u_1(t), u_2(t), ..., u_n(t))$, by a set of n balls, where each ball is constructed independently by one of the nodes. Each node, p_i, constructs a ball $B(e(t), u_i(t))$, which is centered at $\frac{e(t)+u_i(t)}{2}$, and has a radius of $\left\| \frac{e(t)-u_i(t)}{2} \right\|$. Note that at any given time each node has all the information required to independently construct its ball. Theorem 1 states that $\mathrm{Conv}(e(t), u_1(t), u_2(t), ..., u_n(t)) \subset \bigcup_i B(e(t), u_i(t))$.

The application of Theorem 1 is illustrated in Figure 1, which depicts a setup comprised of 5 nodes, each holding a statistics vector, $v_i(t) \in \mathbb{R}^2$. The drift vectors held by the nodes $(u_1(t)..., u_5(t))$, the global statistics vector $v(t)$, and the estimate vector $e(t)$ are depicted, as are the balls constructed by the nodes. The convex hull of the drift vectors is highlighted in gray, and one can see that, as the theorem states, the area defined by the convex hull is bounded by the set of balls.

3.1   Local Constraints

The local constraint on each stream is set as follows: the monitored function f and threshold r can be seen as inducing a coloring over $\mathbb{R}^d$. The vectors

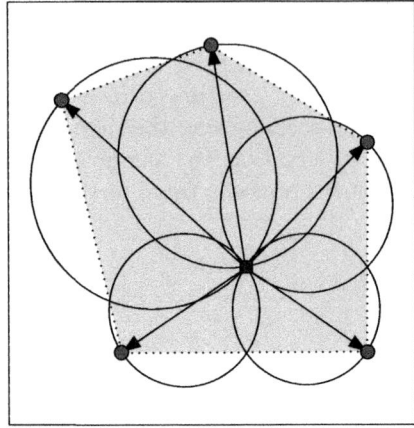

Fig. 1. Illustration of Theorem 1. The drift vectors held by 5 nodes and the balls constructed by them are depicted. The convex hull of the drift vectors is highlighted in gray. As stated by the theorem, the union of the balls bounds the convex hull.

$\{x|f(x) > r\}$ are said to be green, while the vectors $\{y|f(y) \leq r\}$ are said to be red. The local constraint each node maintains is to check whether the ball $B(e(t), u_i(t))$ (the ball centered at $\frac{e(t)+u_i(t)}{2}$, and having a radius of $\left\|\frac{e(t)-u_i(t)}{2}\right\|$) is monochromatic, i.e., whether all the vectors contained in the ball have the same color. Testing for monochromicity is done by finding the maximal and minimal values of f in the ball. This is done locally at each node hence has no effect on the communication load.

If all the local constraints are upheld, the estimate vector is correct: because all the balls contain the estimate vector, and all the balls are monochromatic, the set of vectors defined by the union of all the balls is monochromatic as well. Since the union of all the balls contains the convex hull of the drift vectors and the estimate vector $(\text{Conv}(e(t), u_1(t), u_2(t), ..., u_n(t)))$, and according to Equation (4) the global statistics vector is contained in the convex hull of the drift vectors, the estimate vector and the global statistics vector have the same color. Therefore, they are on the same side of the threshold, i.e., the estimate vector is correct.

4   Distributed Monitoring

In this section we present algorithms that are based on the geometric method for decomposing the monitoring task into local constraints on the streams. After presenting the algorithms, we describe how they can be tuned to relax performance requirements in favour of reducing communication load, and how they can be modified to support time-varying weights.

4.1   The Decentralized Algorithm

Following is a simple, broadcast-based algorithm for monitoring threshold functions: each node maintains a copy of the last statistics vector sent by each of the nodes. The initialization stage consists of every node broadcasting its initial statistics vector. Upon receipt of all the initialization messages, each node calculates the estimate vector, $e(t)$. Then, as more data arrives on the stream, each node can check its local constraint according to the estimate vector and its drift vector. If a local constraint is violated at a node, p_i, the node broadcasts a message of the form $< i, v_i(t) >$, containing its identifier and its local statistics vector at the time. The broadcasting node updates its v'_i parameter and recalculates the estimate vector. Upon receiving a broadcast message from a node, p_i, each node updates its v'_i parameter and recalculates the estimate vector.

  If all the local constraints are upheld, Theorem 1 guarantees the correctness of the estimate vector (enabling every node to locally calculate the value of the threshold function). After a node broadcasts a message, its local constraint is upheld (because the ball it constructs has a radius of 0, and therefore is monochromatic). If a local constraint has been violated, at worst all n nodes (but possibly fewer) will broadcast a message before all the local constraints are upheld again.

  A formal description can be found in Algorithm 1.

Algorithm 1. The decentralized algorithm

Initialization: at a node p_i

- Broadcast a message containing the initial statistics vector and update v'_i to hold the initial statistics vector. Upon receipt of messages from all the nodes, calculate the estimate vector $(e(t))$.

Processing Stage at a node p_i:

- Upon arrival of new data on the local stream, recalculate $v_i(t)$, and $u_i(t)$, and check if $B(e(t), u_i(t))$ remains monochromatic. If not, broadcast the message $<i, v_i(t)>$ and update v'_i to hold $v_i(t)$.
- Upon receipt of a new message $<j, v_j(t)>$, update v'_j to hold $v_j(t)$, recalculate $e(t)$, and check if $B(e(t), u_i(t))$ is monochromatic. If $B(e(t), u_i(t))$ is not monochromatic, broadcast the message $<i, v_i(t)>$ and update v'_i to hold $v_i(t)$.

4.2   The Coordinator-Based Algorithm

Local constraints are also used in the coordinator-based algorithm, but the coordinator is responsible for calculating the estimate vector, maintaining its correctness, and distributing it to the other nodes. In the decentralized algorithm the violation of a constraint on one node requires communicating with all the rest of the nodes (a broadcast message is sent). While this may be a good solution in setups where the nodes are closely coupled, in other cases we can further reduce the communication load by introducing a coordinator. The presence of a coordinator enables us to resolve a violation at a node by communicating with only a subset of the nodes, as opposed to communicating with all the nodes as required in the decentralized algorithm. Consider, for example, a set of equally weighted nodes monitoring the function $f(x) = (x - 5)^2$ (a function over a single dimensional statistics vector), and a threshold value of $r = 9$. Say that at time t the estimate vector is $e(t) = 5$. Note that since $f(e(t)) = 0 < r$, any drift vector in the range $[2, 8]$ satisfies the local constraint at the node. Let us assume that the drift vector at the coordinator, p_1, is $u_1(t) = 4$, and the constraints at all n nodes are satisfied except for p_2, which holds the drift vector $u_2(t) = 1$. In the decentralized algorithm, since the constraint at p_2 has been violated, it would have broadcast its statistics vector to all n nodes. However, the constraint violation at p_2 can be resolved by setting the drift vector at both p_1 and p_2 to the average of the drift vectors on both nodes, i.e., by setting $u_1(t) = u_2(t) = \frac{u_1(t) + u_2(t)}{2} = 2.5$. After this averaging operation, drift vectors on both p_1 and p_2 are within the range $[2, 8]$, and thus all the local constraints are upheld. Note that this action preserves the convexity property of the drift vector (Property 3). The act of averaging out a subset of drift vectors in order to resolve a violated constraint is called a balancing process.

In order to facilitate the balancing of vectors, every node p_i holds a slack vector denoted by δ_i, as defined in Section 2. The slack vector is first normalized by dividing it by the weight assigned to the node. Then it is added to the drift vector as specified in Equation (2). The coordinator is responsible for ensuring that the sum of all slack vectors is $\mathbf{0}$, thus maintaining the convexity property of the drift vectors (Property 3).

To initialize the algorithm, each node sends its initial statistics vector to the coordinator. Initially the slack vector held by each node is set to $\mathbf{0}$. The coordinator calculates the estimate vector and sends it to the rest of the nodes. As more data arrives on a node's stream, the node checks its local constraint. If a local constraint is violated at one of the nodes, it notifies the coordinator by sending it a message containing its current drift vector and its current statistics vector. The coordinator first tries to resolve the constraint violation by executing a balancing process.

During the balancing process the coordinator tries to establish a group of nodes (called the balancing group and denoted by P'), such that the average of the drift vectors held by the nodes in the balancing group (called the balanced vector and denoted by b), creates a monochromatic ball with the estimate vector i.e., such that $B(e(t), b)$ is monochromatic. The balanced vector is calculated as follows:

$$b = \frac{\sum\limits_{p_i \in P'} w_i u_i(t)}{\sum\limits_{p_i \in P'} w_i} \tag{5}$$

The balancing process proceeds as follows: when a node p_i notifies the coordinator that its local constraint has been violated, it appends its drift vector and its current statistics vector to the message. The coordinator constructs a balancing group consisting of p_i and itself. It then checks if the ball defined by the balanced vector, $B(e(t), b)$, is monochromatic. If $B(e(t), b)$ is not monochromatic, the coordinator randomly selects a node that is not in the balancing group, and requests it to send its drift vector and local statistics vector. Then it adds that new node to the balancing group and rechecks $B(e(t), b)$. The process is performed iteratively until either $B(e(t), b)$ is monochromatic, or the balancing group contains all the nodes. If the coordinator established a balancing group such that $B(e(t), b)$ is monochromatic, the balancing process is said to have succeeded. In this case the coordinator sends each node in the balancing group an adjustment to its slack vector. This causes the drift vectors held by all nodes in the balancing group to be equal to b. The adjustment to the slack vector sent to each node $p_i \in P'$ is denoted by $\Delta\delta_i$, and is calculated as follows:

$$\Delta\delta_i = w_i b - w_i u_i(t)$$

After receiving the slack vector adjustment, each node simply adds the adjustment to the current value, i.e., $\delta_i \leftarrow \delta_i + \Delta\delta_i$. One can easily verify that after a successful balancing process the sum of all slack vectors remains 0, and the drift vector held by each node in the balancing group is b, thus resolving the original constraint violation.

If the balancing process has failed (i.e., the balancing group contains all the nodes, and $B(e(t), b)$ is not monochromatic), the coordinator calculates a new estimate vector (according to the updated statistics vectors sent by the nodes) and sends it to all the nodes. Upon receipt of the new estimate vector, the nodes set their slack vectors to 0 and modify their v'_i parameter to hold the value of the statistics vector they sent to the coordinator during the balancing process, thus resolving the original constraint violation.

In order to implement the algorithm, the following messages must be defined:

<INIT,v_i> Used by nodes to report their initial statistics vector to the coordinator in the initialization stage.

<REQ>   Used by the coordinator during the balancing process to request that a node send its statistics vector and drift vector.

<REP,v_i,u_i> Used by nodes to report information to the coordinator when a local constraint has been violated, or when the coordinator requests information from the node.

<ADJ-SLK,$\Delta\delta_i$> Used by the coordinator to report slack vector adjustments to nodes after a successful balancing process.

<NEW-EST,e> Used by the coordinator to report to the nodes a new estimate
vector.

A formal description is given in Algorithm 2.

4.3   Relaxing the Precision Requirements

A desired trade-off when monitoring threshold functions is between accuracy and
communication load. In some cases an approximate value of the threshold function is sufficient, that is, the correct value of the threshold function is required
only if the value of the monitored function is significantly far from the threshold.
In other words, if ε is a predetermined error margin, and if $f(v(t)) > r + \varepsilon$ or
$f(v(t)) \leq r - \varepsilon$, we require that the estimate vector, $e(t)$, be correct, but we do
not require it if $r - \varepsilon < f(v(t)) \leq r + \varepsilon$.

Consider the feature monitoring example given in Section 1. Say we would
like to select all the features whose information gain is above 0.05. Obviously, it
is important to select a feature whose information gain is significantly high, and
not to select a feature whose information gain is significantly low. For example,
it is important to select a feature whose information gain score is 0.1, and not
to select a feature whose information gain score is 0.01. Including or excluding
features whose information gain score is very close to the threshold value, for
example a feature whose information gain score is 0.048, will probably not have
a significant effect on the quality of the selected feature set, while the cost of
monitoring such features is expected to be high, since their information gain is
expected to fluctuate around the threshold value. Therefore we can significantly
improve the efficiency of our monitoring algorithms if we set some error margin,
say 0.005. In other words, features that are currently selected will be removed
from the set of selected features only when their information gain falls below
0.045, and features that are currently not selected will be added to the set of
selected feature only if their information gain rises above 0.055.

Our algorithm can be easily tuned to relax the precision requirements by an
error margin of ε as follows: instead of working with a single coloring, induced
by the monitored function f and the threshold value r, two sets of coloring are
defined, one induced by the monitored function f and the threshold value $r + \varepsilon$,
and a second induced by the monitored function f and the threshold value $r - \varepsilon$.
Whenever the original algorithm checks whether a ball is monochromatic, then, if
$f(v(t)) \leq r$, the modified algorithm will check whether the ball is monochromatic
according to the first coloring (the one induced by f and $r + \varepsilon$). If $f(v(t)) > r$,
the modified algorithm will check whether the ball is monochromatic according
to the second coloring (the one induced by f and $r - \varepsilon$). This ensures that if all
the balls are in the range defined by $\{x | r - \varepsilon < f(x) \leq r + \varepsilon\}$, no messages are
transmitted.

4.4   Handling Time-Varying Weights

Up to this point we have assumed that the weights assigned to nodes are fixed,
such as when the weights are the size of sliding windows used for collecting data

---

Algorithm 2. The coordinator-based algorithm

Initialization:

- Send an INIT message to the coordinator, set v' to hold the initial statistics vector, and set the slack vector to 0. Upon receipt of messages from all nodes, the coordinator calculates the estimate vector and informs the nodes via a NEW-EST message.

Processing Stage at an Ordinary Node p_i:

- Upon arrival of new data on a node's local stream, recalculate $v_i(t)$ and $u_i(t)$, and check if $B(e(t), u_i(t))$ remains monochromatic. If not, send a $<$REP,$v_i(t)$,$u_i(t)>$ message to the coordinator, and wait for either a NEW-EST or an ADJ-SLK message.
- Upon receipt of a REQ message, send a $<$REP,$v_i(t)$,$u_i(t)>$ message to the coordinator and wait for either a NEW-EST or ADJ-SLK message.
- Upon receipt of a NEW-EST message, update the estimate vector $(e(t))$ to the value specified in the message, set the value of v' to the statistics vector sent to the coordinator, and set the slack vector to 0.
- Upon receipt of an ADJ-SLK message, add the value specified in the message to the value of the slack vector $(\delta_i \leftarrow \delta_i + \Delta\delta_i)$.

Processing Stage at the Coordinator:

- Upon arrival of new data on the local stream, recalculate $v_1(t)$ and $u_1(t)$, and check if $B(e(t), u_1(t))$ remains monochromatic. If not, initiate a balancing process, setting the balancing group to
$P' = \{< 1, v_1(t), u_1(t) >\}$.
- Upon receipt of a REP message from the node p_i, initiate a balancing process, setting the balancing group to
$P' = \{< 1, v_1(t), u_1(t) >, < i, v_i, u_i >\}$.

Balancing Process at the Coordinator:

1. Calculate balanced vector, b, according to Equation (5). If the ball $B(e(t), b)$ is monochromatic goto (2), otherwise goto (3).
2. For each item in the balancing group, $< i, v_i, u_i >$, calculate the slack vector adjustment, $\Delta\delta_i = w_i b - w_i u_i(t)$, send p_i a $<$ADJ-SLK,$\Delta\delta_i>$ message, and then exit the Balancing Process.
3. If there are nodes not contained in the balancing group, select one of these nodes at random, and send it a REQ message. Upon receipt of the REP message, add the node to the Balancing Group and goto (1). Otherwise calculate a new estimate vector (based on the v_i values received from all the nodes), send a NEW-EST message to all nodes, and exit the Balancing Process.

---

from the streams. We now address cases where weights assigned to nodes may vary with time, as when a node's weight at a given time is the number of data items received on its stream so far.

We next describe the required modifications to the algorithms in order to ensure their correctness in a setup where weights vary with time. In such a setup we denote the weight assigned to the node p_i at time t by $w_i(t)$. Each message in the original algorithms is modified by appending $w_i(t)$ to it. Along with the last vector broadcast by each of the other nodes (v'_i), the nodes in the decentralized algorithm keep track of the last broadcast weight, denoted by w'_i. Nodes calculate $e_i(t)$, $\Delta v_i(t)$, and $u_i(t)$ as follows:

$$e_i(t) = \frac{\sum_{i=1}^{n} w'_i v'_i}{\sum_{i=1}^{n} w'_i}$$

$$\Delta v_i(t) = \frac{w_i(t) v_i(t) - w'_i v'_i(t) - (w_i(t) - w'_i) e_i(t)}{w_i(t)}$$

In the decentralized algorithm the drift vector is calculated by

$$u_i(t) = e_i(t) + \Delta v_i(t)$$

and in the coordinator-based algorithm, by

$$u_i(t) = e(t) + \Delta v_i(t) + \frac{\delta_i}{w_i(t)}$$

In the coordinator-based algorithm, the balanced vector and the slack vector adjustments are calculated according to the weights appended to the messages:

$$b = \frac{\sum_{i=1}^{k} w_i(t) u_i(t)}{\sum_{i=1}^{k} w_i(t)}$$

$$\Delta \delta_i = w_i(t) b - w_i(t) u_i(t)$$

Note that if the weights are fixed, $e_i(t)$, $\Delta v_i(t)$, and $u_i(t)$ hold the same values they hold in the original algorithms. Furthermore, one can easily verify that the new definitions of these parameters maintain Equation (3) i.e.,

$$v(t) = \frac{\sum_{i=1}^{n} w_i(t) u_i(t)}{\sum_{i=1}^{n} w_i(t)}, \text{ and thus maintain the correctness of the algorithm.}$$

5   Performance Analysis

We would like to determine how the various parameters of the monitoring problem affect the communication load generated by the proposed algorithms. In order to do so we present a simplified model of our algorithm and analyze the probability that a constraint violation will occur at a node. Since a constraint violation is the trigger for communications in both algorithms, this analysis should provide indications regarding the generated communication load.

Generally speaking, the dominant factor affecting the performance of the algorithms is the average distance of the estimate vector from the set of vectors for which the value of the monitoring function equals the threshold value. More formally, let the threshold set defined by the monitoring function f and threshold value r be the set of vectors for which the value of the threshold function equals the threshold value. Let the threshold set be denoted by $T(f, r)$, i.e.,

$$T(f, r) = \{ \boldsymbol{x} | f(\boldsymbol{x}) = r \}$$

Let the distance of a vector $\boldsymbol{x}$ from the threshold set $T(f, r)$, denoted as $dist(\boldsymbol{x}, f, r)$, be the minimum distance of $\boldsymbol{x}$ from any point in $T(f, r)$ i.e.,

$$dist(\boldsymbol{x}, f, r) = \min(\| \boldsymbol{y} - \boldsymbol{x} \| \, | f(\boldsymbol{y}) = r)$$

The farther the estimate vector is, at a given time, from the threshold set, the more the local statistics vectors can change without violating local constraints. Therefore, a greater average distance of the estimate vector from the threshold set will result in a greater reduction in communications.

The average distance of the estimate vector from the threshold set is affected by many parameters. To begin with, it is affected by the coloring induced by the monitored function and the threshold value. Figure 2 illustrates the coloring induced by two sets of a monitored function and a threshold value. Figure 2(a) illustrates the coloring induced by the function $f_1(x, y) = \sin(2\sqrt{(x^2 + y^2)})$ and the threshold value 0, and Figure 2(b) illustrates the coloring induced by the function $f_2(x, y) = \frac{1}{1+e^{-x}} + \frac{1}{1+e^{-y}}$ and the threshold value 0.75 ($f_2(x, y)$ is a simple two-layer neural net).

It is clear that the distance of any point in $\mathbb{R}^2$ from the threshold set defined by f_1 and 0 cannot be greater than $\frac{\pi}{4} \approx 0.785$. Therefore, the average distance of the estimate vector from the threshold set in this case is bounded from above by 0.785, thus yielding a relatively low reduction in communications when monitored by our algorithms. However, the maximum distance of a point in $\mathbb{R}^2$ from the threshold set defined by f_2 and 0.75 is unbounded, and the dominating factor affecting the performance of our algorithms in this case is the nature of the data received on the streams.

In order to analyze how our algorithms are affected by the nature of the data on the streams, we consider periods during which this data is stationary (this fact, however, is not known to any of the nodes). More formally, we assume that each stream item is a d-dimensional vector, where the jth component is independently drawn from a random variable denoted by X_j with a defined expectancy and variance, denoted by $E[X_j]$ and $V[X_j]$ respectively. We assume the system consists of n nodes, and that each node holds a sliding window of N items.

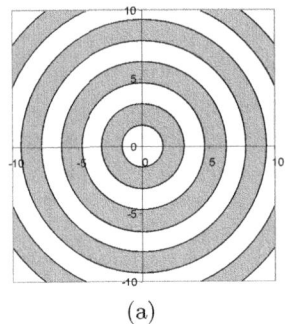
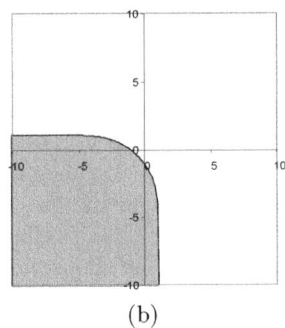

(a)   (b)

Fig. 2. The colorings induced by two sets of monitored functions and threshold values. (a) depicts the coloring induced by $f_1 = \sin(2\sqrt{x^2+y^2}) \geq 0$, and (b) depicts the coloring induced by $f_2 = \frac{1}{1+e^{-x}} + \frac{1}{1+e^{-y}} \geq 0.75$.

We denote the last N items received on the stream monitored by p_i as $v_{i,1}, v_{i,2}, ..., v_{i,N}$, and the components of a vector as follows: $v_{i,k} = (v_{i,k}^{(1)}, v_{i,k}^{(2)}, ..., v_{i,k}^{(d)})$. The local statistics vector held by a node is the average of the items contained in its sliding window, and the global statistics vector is the average of the items contained in the sliding windows held by all the nodes, i.e.,

$$v_i(t) = \frac{1}{N} \sum_{k=1}^{N} v_{i,k} \; ; \; v(t) = \frac{\sum_{i=1}^{n} \sum_{k=1}^{N} v_{i,k}}{N \cdot n}$$

It is easy to see that the expected value for the global statistics vector and each local statistics vector is $E[v(t)] = E[v_i(t)] = (E[X_1], E[X_2], ..., E[X_d])$.

Figure 3 depicts the coloring induced by f_2 and the threshold value 0.75, the expected global statistics vector, and the distance of the expected global statistics vector from the threshold set. The expected global statistics vector and its distance from the threshold set define a sphere called the distance sphere. We denote the distance of the expected global statistics vector from the threshold set by D_{global} i.e., $D_{global} = dist(E[v(t)], f, r)$.

We present the following simplified model of our algorithms: we assume that the estimate vector holds the value of the expected global statistics vector, i.e., $e(t) = E[v(t)] = (E[X_1], E[X_2], ..., E[X_d])$. Furthermore, we assume that data on each stream arrives in blocks of N items. We would like to bound $\Pr_{violation}$, the probability that the arrival of a new block of data items on a stream will cause a constraint violation at the node monitoring the stream.

We assume that the estimate vector is the expected global statistics vector. Consequently, as long as the local statistics vector held by a node is contained within the distance sphere, the constraint checked by the node is guaranteed not to be violated. That is, if $\|E[v(t)] - v_i(t)\| < D_{global}$, then $B(E[v(t)], v_i(t))$ is fully contained in the distance sphere (see Figure 3).

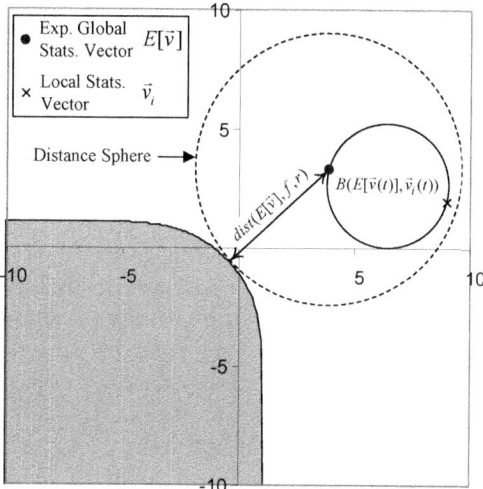

Fig. 3. Depicts the coloring induced by the function $f_2 = \frac{1}{1+e^{-x}} + \frac{1}{1+e^{-y}}$ and the threshold value 0.75, together with the expected global statistics vector, its distance from the threshold set, and a local statistics vector, that is contained within the distance sphere. One can see that the ball $B(E[v(t)], v_i(t))$ is fully contained in the distance sphere.

Therefore, the probability that a constraint violation will occur at a node is less than the probability that the local statistics vector held by the node will not be contained in the distance sphere. Using the Markov Inequality we obtain:

$$\Pr_{violation} \leq \frac{\sum\limits_{i=1}^{d} V[X_i]}{N \cdot (D_{global})^2}$$

If the components of the data vectors are bounded between 0 and 1 – as happens in the important case in which they represent probabilities of terms to appear in a document – the Hoeffding bound can be used, to show that:

$$\Pr_{violation} \leq \exp\left(-2\left(D_{global}^2 - \frac{\sum\limits_{i=1}^{d} V[X_i]}{N}\right)^2 / d\right)$$

Both bounds decrease quickly when D_{global} increases. This suggests that for data mining applications, features with a small information gain will not cause many constraint violations at any node, since their D_{global} is large. This is practically important, since usually most of the candidate features have a rather small information gain, and thus the proposed algorithm will considerably reduce communication. This is supported by the experimental results presented in the next section.

6 Experimental Results

We performed several experiments with the decentralized algorithm. We tested the algorithm in a distributed feature selection setup. We used the Reuters Corpus (RCV1-v2) [12] in order to generate a set of data streams. RCV1-v2 consists of 804414 news stories, produced by Reuters between August 20, 1996, and August 19, 1997. Each news story, which we refer to as a document, has been categorized according to its content, and identified by a unique document id.

RCV1-v2 has been processed by Lewis, Yank, Rose, and Li [12]. Features were extracted from the documents, and indexed. A total of 47236 features were extracted. Each document is represented as a vector of the features it contains. We refer to these vectors as feature vectors. We simulate n streams by arranging the feature vectors in ascending order (according to their document id), and selecting feature vectors for the streams in a round robin fashion.

In the original corpus each document may be labeled as belonging to several categories. The most frequent category documents are labeled with is "CCAT" (the "CORPORATE/INDUSTRIAL" category). In the experiments our goal is to select features that are most relevant to the "CCAT" category, therefore each vector is labeled as positive if it is categorized as belonging to "CCAT", and negative otherwise.

Unless specified otherwise, each experiment was performed with 10 nodes, where each node holds a sliding window containing the last 6700 documents it received. In each experiment we used the decentralized algorithm in order to detect for each feature, at any given time, whether its information gain in above or below a given threshold value. At any given time the information gain of a feature is based on the documents contained at the time in the sliding windows of all the nodes.

The experiments were designed to explore several properties of the algorithm. We were interested in determining how various parameters of the monitoring task affect the performance of the algorithm. The parameters of the monitoring task can be divided into characteristics of the monitoring task, and tunable parameters. The characteristics of the monitoring task include the number of streams to be monitored, and the desired threshold value. Tunable parameters include the size of the sliding window used by each node, and the permitted error margin. In addition we were interested in examining the behaviour of the algorithm when used for simultaneously monitoring several features.

In order to examine the effect of the various parameters on the performance of the algorithm, we chose three features that display different characteristic behaviour. The chosen features are "bosnia", "ipo", and "febru". Figure 4 depicts how the information gain for each feature evolves over the streams. The information gain for the feature "bosnia" displays a declining trend as the stream

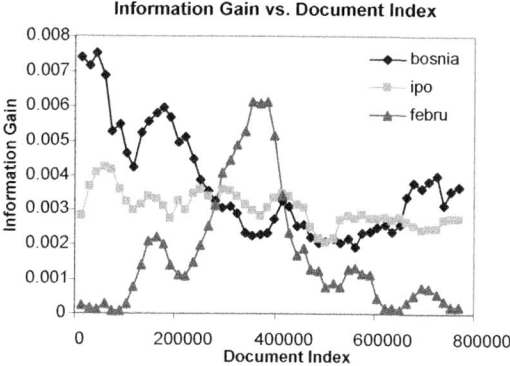

Fig. 4. Information gain for the features "bosnia", "ipo", and "febru" as it evolves over the streams. The information gain for the feature "bosnia" displays a declining trend as the stream evolves. The information gain for the feature "ipo" remains relatively steady, while the information gain for the feature "febru" peaks about halfway through the stream.

evolve. The information gain for the feature "ipo" remains relatively steady, while the information gain for the feature "febru" peaks about halfway through the stream.

We start by examining the influence of the characteristics of the monitoring task on the performance of the algorithm. Figure 5 show the number of broadcasts produced when monitoring each one of the features for threshold values ranging from 0.00025 to 0.006. In addition the cost incurred by the naive algorithm is plotted, i.e., the number of messages required for collecting all the data to a central location. One can notice that even for adverse threshold values, the algorithm incurs a significantly lower communication cost than the cost incurred by the naive algorithm.

In order to check the effect the number of nodes has on the performance of the algorithm, we performed the following experiment: the stream of documents was divided in advance into 100 sub-streams in a round robin fashion. Simulations were run with the number of nodes ranging from 10 to 100. In a simulation consisting of n nodes, the first n sub-streams were used. This methodology ensures that the characteristics of the streams remain similar when simulating different numbers of nodes. Each node held a sliding window of 670 items.

Obviously, increasing the number of nodes will increase the number of broadcasts required in order to perform the monitoring task. Since the nodes in our experiment receive streams with similar characteristics, we expect that the number of broadcasts will increase linearly.

Two sets of simulation were run, the first with a threshold value of 0.003, and the second with a threshold value of 0.006. The results are plotted in Figure 6. Both graphs show that the number of broadcasts increases linearly as more nodes

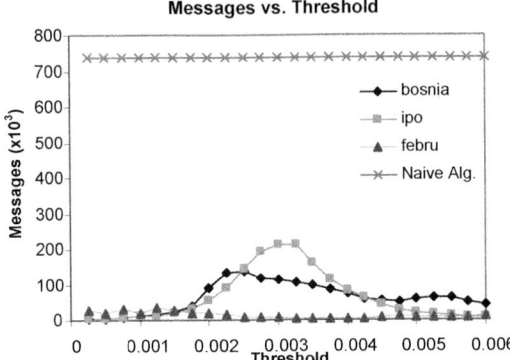

Fig. 5. Number of broadcasts produced in order to monitor each feature as a function of the threshold value. In addition, the cost incurred by monitoring a feature by a naive algorithm is plotted. Even for adverse threshold values our algorithm performs significantly better than the naive algorithm.

are added. Comparing the two graphs reveals that the number of broadcasts increases more moderately when using a threshold value of 0.006. This is due to the fact that as indicated in Figure 4, the average information gain on the monitored features is closer to 0.003.

Next we performed two experiments in order to evaluate the effect of tunable parameters on the performance of the algorithm. We performed the following experiments on the three features: for each feature we chose the threshold value that incurred the highest communication cost (0.0025 for "bosnia", 0.003 for "ipo", and 0.00125 for "febru"). We ran a set of simulations on each feature, using error margin values ranging from 0 to 50 percent of the threshold value. Then we ran an additional set of simulations for each feature, setting the size of the sliding window used by each node to values ranging from 6700 items to 13400 items. The results of these experiments are plotted in Figure 7. The results indicate that increasing the error margin is very effective in reducing the communication load. Using an error margin as small as 5 percent significantly reduces the communication load. Increasing the window size also reduces the communications load. The effect of increasing the window size is most evident for the feature "ipo", which incurs the highest communication cost among the three features. In general, increasing the window size has a greater effect the closer the information gain of feature is to the threshold value.

Finally, we checked the performance of the algorithm when simultaneously monitoring multiple features. As the number of features that are monitored simultaneously increases, the probability that a constraint on one of the features will be violated when a new data item is received increases as well. Furthermore, a constraint violation can cause a cascading effect. A constraint violation for a feature at one of the nodes causes all the nodes to calculate a new estimate

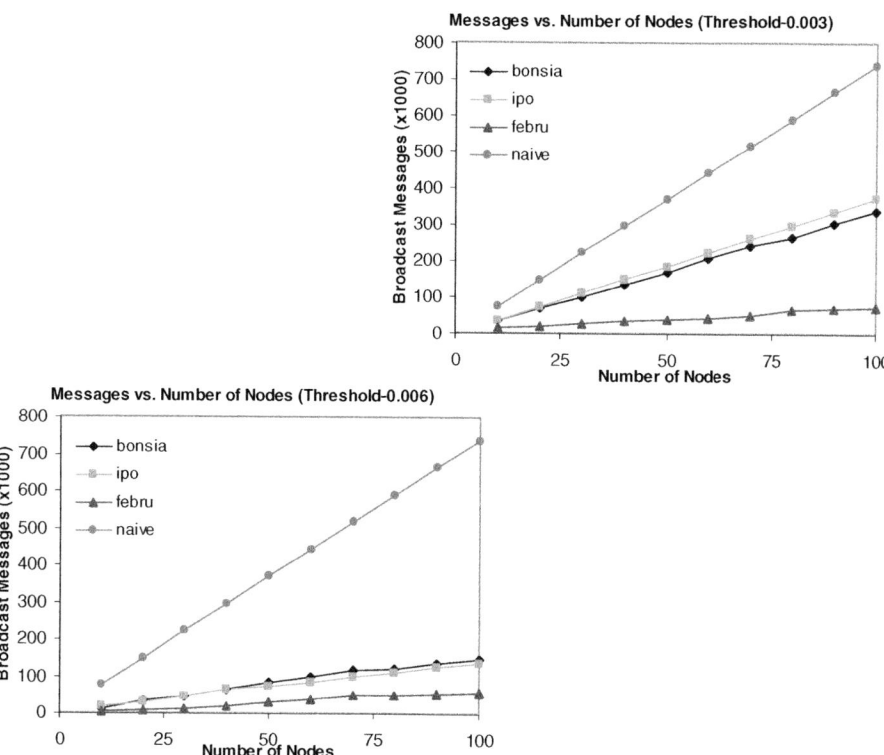

Fig. 6. Number of messages produced in relation to the number of nodes. The number of messages increases linearly as the number of nodes increases, indicating that the algorithm scales well. Since the average information gain of all the features is closer to 0.003 than to 0.006, the number of messages increases more moderately when using a threshold value of 0.006.

vector for the feature. Since the value of the estimate vector for the feature has changed, the constraint for the feature may be violated at additional nodes, causing these nodes to broadcast. The purpose of this experiment is to determine the number of simultaneous features the algorithm can monitor while remaining efficient, i.e., incurring a cost that is lower than the cost incurred by the naive algorithm. The experiment consisted of a series of simulations, using a threshold value of 0.001. In each simulation a number of features were selected randomly. Simulations were run with the number of features ranging from 1 to 5000. The results of this experiment are plotted in Figure 8. In addition the cost incurred by the naive algorithm is plotted.

The results indicate the algorithm remains efficient when simultaneously monitoring several thousands of features, but is inefficient when simultaneously monitoring more than about 4500 features.

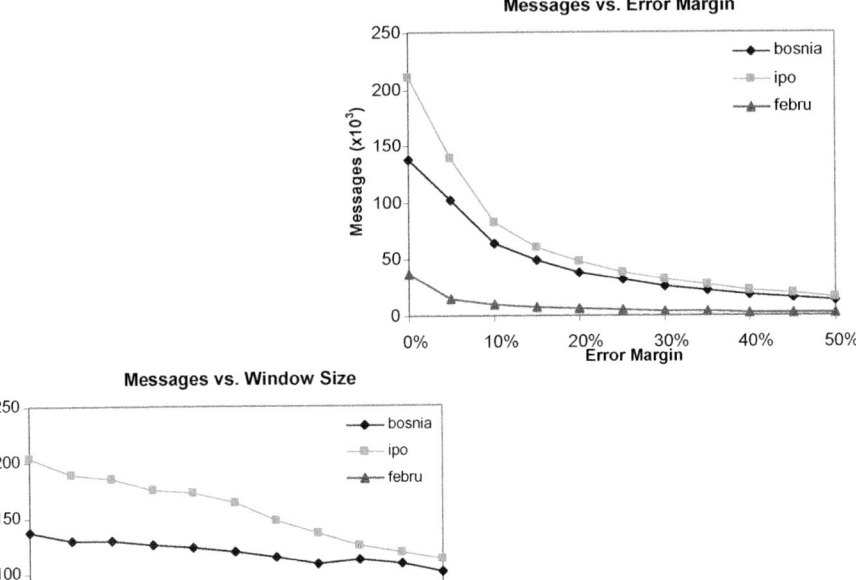

Fig. 7. The influence of tunable parameters on performance. Increasing the error margin is more effective in reducing the communication load than increasing the window size. Using an error margin as small as 5 percent significantly reduces the communication load.

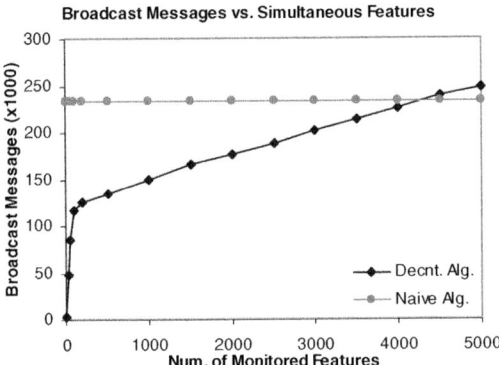

Fig. 8. Number of messages in relation to the number of simultaneously monitored features. Our algorithm remains efficient when simultaneously monitoring up to about 4500 features.

7   Conclusion

Monitoring streams over distributed systems is an important challenge which has a wide range of applications. Scalability and efficiency of proposed solutions strongly depend on the volume and frequency of communication operations. However, despite the amount of work that was invested in this direction, most of the efficient solutions found in the literature can only be applied to simple aggregations or to linear functions. Most probably the reason is that when the function is non linear, effects seen in one – or only a few – of the streams may often turn out to be misleading with regards to the global picture.

The described solution is a general framework for monitoring arbitrary threshold functions over a system of distributed streams. The evaluation of this approach using real-life data, applied to the information gain function, reveals that it is highly effective in reducing communication frequency.

References

1. Online data mining for co-evolving time sequences. In: ICDE 2000: Proceedings of the 16th International Conference on Data Engineering, Washington, DC, USA, p. 13. IEEE Computer Society, Los Alamitos (2000)
2. Fjording the stream: An architecture for queries over streaming sensor data. In: ICDE 2002: Proceedings of the 18th International Conference on Data Engineering (ICDE 2002), Washington, DC, USA, p. 555. IEEE Computer Society, Los Alamitos (2002)
3. Alon, N., Matias, Y., Szegedy, M.: The space complexity of approximating the frequency moments. In: STOC 1996: Proceedings of the Twenty-Eighth Annual ACM Symposium on Theory of Computing, pp. 20–29. ACM Press, New York (1996)
4. Babcock, B., Babu, S., Datar, M., Motwani, R., Widom, J.: Models and issues in data stream systems. In: PODS 2002: Proceedings of the Twenty-First ACM SIGMOD-SIGACT-SIGART Symposium on Principles of Database Systems, pp. 1–16. ACM Press, New York (2002)
5. Babu, S., Widom, J.: Continuous queries over data streams. SIGMOD Rec. 30(3), 109–120 (2001)
6. Bulut, A., Singh, A.K., Vitenberg, R.: Distributed data streams indexing using content-based routing paradigm. In: IPDPS. IEEE Computer Society, Los Alamitos (2005)
7. Carney, D., Çetintemel, U., Cherniack, M., Convey, C., Lee, S., Seidman, G., Stonebraker, M., Tatbul, N., Zdonik, S.B.: Monitoring streams - a new class of data management applications. In: VLDB, pp. 215–226 (2002)
8. Cherniack, M., Balakrishnan, H., Balazinska, M., Carney, D., Cetintemel, U., Xing, Y., Zdonik, S.: Scalable Distributed Stream Processing. In: CIDR 2003 - First Biennial Conference on Innovative Data Systems Research, Asilomar, CA (January 2003)
9. Motwani, R., Widom, J., Arasu, A., Babcock, B., Babu, S., Datar, M., Manku, G., Olston, C., Rosenstein, J., Varma, R.: Query processing, resource management, and approximation in a data stream management system. In: CIDR 2003 - First Biennial Conference on Innovative Data Systems Research, Asilomar, CA, pp. 245–256 (2003)

10. Cormode, G., Garofalakis, M., Muthukrishnan, S., Rastogi, R.: Holistic aggregates in a networked world: distributed tracking of approximate quantiles. In: SIGMOD 2005: Proceedings of the 2005 ACM SIGMOD International Conference on Management of Data, pp. 25–36 (2005)

11. Dilman, M., Raz, D.: Efficient reactive monitoring. In: INFOCOM, pp. 1012–1019 (2001)

12. Lewis, D.D., Yang, Y., Rose, T.G., Li, F.: RCV1: A New Benchmark Collection for Text Categorization Research. Journal of Machine Learning Research, 361–397 (2004)

13. Liu, L., Pu, C., Tang, W.: Continual queries for internet scale event-driven information delivery. IEEE Transactions on Knowledge and Data Engineering 11(4), 610–628 (1999)

14. Madden, S., Shah, M., Hellerstein, J.M., Raman, V.: Continuously adaptive continuous queries over streams. In: SIGMOD 2002: Proceedings of the 2002 ACM SIGMOD International Conference on Management of Data, pp. 49–60. ACM Press, New York (2002)

15. Manjhi, A., Shkapenyuk, V., Dhamdhere, K., Olston, C.: Finding (recently) frequent items in distributed data streams. In: ICDE 2005: Proceedings of the 21st International Conference on Data Engineering (ICDE 2005), Washington, DC, USA, pp. 767–778. IEEE Computer Society, Los Alamitos (2005)

16. Manku, G.S., Motwani, R.: Approximate frequency counts over data streams. In: VLDB, pp. 346–357 (2002)

17. Terry, D., Goldberg, D., Nichols, D., Oki, B.: Continuous queries over append-only databases. In: SIGMOD 1992: Proceedings of the 1992 ACM SIGMOD International Conference on Management of Data, pp. 321–330. ACM Press, New York (1992)

18. Zhu, Y., Shasha, D.: Statstream: Statistical monitoring of thousands of data streams in real time. In: VLDB, pp. 358–369 (2002)

Privacy Preserving Spatio-temporal Clustering on Horizontally Partitioned Data*

Ali Inan[1] and Yucel Saygin[2]

[1] Department of Computer Science,
The University of Texas at Dallas
Richardson, USA
[2] Faculty of Engineering and Natural Sciences
Sabanci University,
Istanbul, Turkey
ysaygin@sabanciuniv.edu

Abstract. Space and time are two important features of data collected in ubiquitous environments. Such time-stamped location information is regarded as spatio-temporal data and, by its nature, spatio-temporal data sets, when they describe the movement behavior of individuals, are highly privacy sensitive. In this chapter, we propose a privacy preserving spatio-temporal clustering method for horizontally partitioned data. Our methods are based on building the dissimilarity matrix through a series of secure multi-party trajectory comparisons managed by a third party. Our trajectory comparison protocol complies with most trajectory comparison functions. A complexity analysis of our methods shows that our protocol does not introduce extra overhead when constructing dissimilarity matrices, compared to the centralized approach.

1 Introduction

Advances in wireless communication technologies resulted in a rapid increase in usage of mobile devices. PDAs, mobile phones and various other devices equipped with GPS technology are now a part of our daily life. One direct consequence of this change is that, using such devices, locations of individuals can be tracked by wireless service providers. Individuals sometimes voluntarily pay for being tracked by means of Location Based Services (LBS) such as vehicle telematics that offer vehicle tracking and satellite navigation. Tracking is also enforced by law in some countries, as in the case of the Enhanced-911 mandate, passed by U.S. Federal Communications Commission in 1996. The mandate requires that any cellular phone calling 911, the nationwide emergency service number, be located within at least 50 to 100 meters.

Time-stamped location information is regarded as spatio-temporal data due to its time and space dimensions and, by its nature, is highly vulnerable to

---

* This work was funded by the Information Society Technologies programme of the European Commission, Future and Emerging Technologies under IST-014915 GeoP-KDD project.

M. May and L. Saitta (Eds.): Ubiquitous Knowledge Discovery, LNAI 6202, pp. 187–198, 2010.

misuse. In fact, privacy issues related to collection, use and distribution of individuals' location information is the main obstacle against extensive deployment of LBSs. Suppressing identifiers from the data does not suffice since trajectories can easily be re-bound to individuals using publicly available information such as home and work addresses. Therefore new privacy preserving knowledge discovery methods, designed specifically to handle spatio-temporal data, are required. Existing privacy preserving data mining techniques are not suitable for this purpose since time-stamped location observations of an object are not plain, independent attributes of this object.

In this work, we propose a privacy preserving clustering technique for horizontally partitioned spatio-temporal data where each horizontal partition contains trajectories of distinct moving objects collected by a separate site. Consider the following scenario where the proposed techniques are applicable: In order to solve traffic congestion, traffic control offices want to cluster trajectories of users. However, the required spatio-temporal data is not readily available but can be collected from GSM operators. GSM operators are not eager to share their data due to privacy concerns. The solution is running a privacy preserving spatio-temporal clustering algorithm for horizontally partitioned data.

Our method is based on constructing the dissimilarity matrix of object trajectories in a privacy preserving manner which can then be input to any hierarchical clustering algorithm. Main contributions are the introduction of a protocol for secure multiparty computation of trajectory distances and its application to privacy preserving clustering of spatiotemporal data. We also provide complexity and privacy analysis of the proposed method.

In Section 2, we provide related work in the area and then formally define the problem in Section 3. Classification of trajectory comparison functions is provided in Section 4. Communication and computation phases of our method are explained in Sections 5 and 6 respectively. We provide complexity and privacy analysis in Section 7 and finally conclude in Section 8.

2 Related Work

Privacy preserving data mining has become a popular research area in the past years. The aim of privacy preserving data mining is ensuring individual privacy while maintaining the efficacy of data mining techniques. Agrawal and Srikant initiated research on privacy preserving data mining with their seminal paper on constructing classification models while preserving privacy [3]. Saygin et al. propose methods for hiding sensitive association rules before releasing the data [14]. Privacy preserving data mining methods can be classified under two headings: data sanitization and secure multi-party computation. Data sanitization approaches sacrifice accuracy for increased privacy, while secure multi-party computation approaches try to achieve both accuracy and privacy at the expense of high communication and computation costs.

Researchers developed methods for privacy preserving clustering. Most of these methods are based on sanitizing the input and they address only centralized data. Merugu and Ghosh propose methods for constructing data mining models from the input data. These models are not considered private information. The overall clustering schema is constructed by merging these models coming from vertically or horizontally distributed data sources [12]. Oliveira and Zaiane propose methods for preserving privacy by reducing the dimensionality of the data [13]. Their method is not applicable to horizontally partitioned data and moreover, results in loss of accuracy. Vaidya and Clifton propose a secure multi-party computation protocol for k-means clustering on vertically partitioned data[15]. Jha et al. [11] propose a privacy preserving, distributed k-means protocol on horizontally partitioned data through secure multi-party computation of cluster means. Inan et al. propose another privacy preserving clustering algorithm over horizontally partitioned data that can handle numeric, categorical and alphanumeric data [10].

Privacy of spatio-temporal data is of utmost importance for individuals since such data is highly vulnerable to misuse. In this work, we focus on spatio-temporal data and propose a secure multi-party comparison protocol that is applicable to most trajectory comparison functions. Previous work on ensuring individual privacy for spatio-temporal data is limited to sanitization approaches and access control mechanisms. Gruteser and Hoh propose confusing paths to garble trajectories of individuals [9]. Beresord and Stajano introduce "mix zones", in which identification of users is blocked and pseudonyms of incoming user trajectories are mixed up while leaving these mixed zones [4]. A detailed discussion on privacy mechanisms through access control and anonymization can be found in [5]. To the best of our knowledge, this work is the first to introduce a secure multi-party solution to privacy problems in spatio-temporal data without any loss of accuracy.

3   Problem Formulation

Spatio-temporal knowledge discovery deals with time-stamped location observations of moving objects. In some applications spatial component may interpreted in a different way. For example, in stock market analysis, the trajectory of a stock is the one-dimensional vector of price fluctuations in time. In weather forecasting, observations are two dimensional measurements of atmospheric pressure and temperature at weather stations. In this chapter, we primarily focus on moving objects and assume that location information is two dimensional as in the case of GPS, neglecting the altitude.

Trajectory T of a moving object X is a set of location observations in the form $O = (t, d)$ where t represents the time dimension and d represents the two dimensional location information. The number of observations for this trajectory is denoted as $length(X)$ and the ith element of T_X is denoted by $T_X(i)$. Figure 1 depicts these notions for the sample one dimensional spatio-temporal data provided in Table 1.

Table 1. Spatio-temporal data for trajectories X and Y

X	Time				
	1	4	7	10	16
Location	2.3	4.5	6.7	3	2
Y	Time				
	2	4	6	8	
Location	4.3	3.6	7	3	

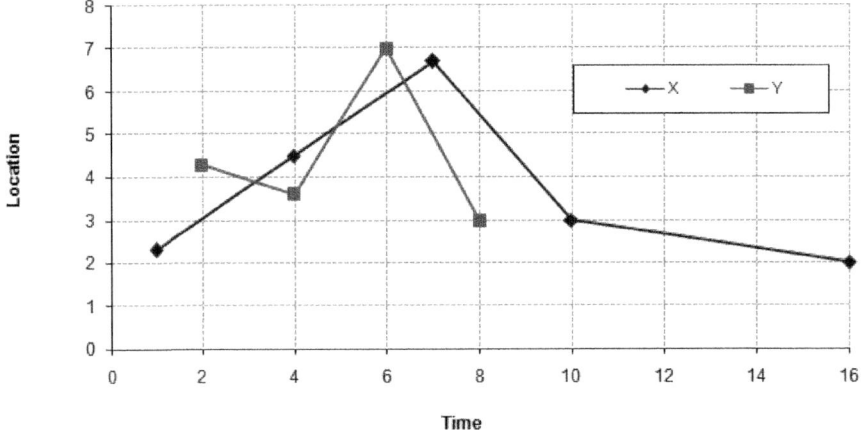

Fig. 1. Trajectories X and Y. $length(X) = 5$ and $length(Y) = 4$

Suppose that there are K data holders, such that $K \geq 2$, which track locations of with unique object id's. The number of objects in data holder k's database is denoted as $size_K$. Data holders want to cluster the trajectories of moving objects without publishing sensitive location information so that clustering results will be public to each data holder at the end of the protocol. There is a distinct third party, denoted as TP, who serves as a means of computation power and storage space. TP's role in the protocol is: (1) managing the communication between data holders, (2) privately constructing the global dissimilarity matrix, (3) clustering the trajectories using the dissimilarity matrix, and (4) publishing the results to the data holders.

Involved parties, including the third party, are assumed to be semi-honest which means that they follow the protocol as they are expected to, but may store any information that is available in order to infer private data in the future. Semi-trusted behavior is also called honest-but-curious behavior. Another assumption is that, all parties are non-colluding, i.e. they do not share private information with each other.

4   Trajectory Comparison Functions

Clustering is the process of grouping similar objects together. In order to measure the similarity between object trajectories, robust comparison functions are needed. However, trajectory comparison is not an easy task since spatio-temporal data is usually collected through sensors and therefore is subject to diverse sources of noise. Under ideal circumstances, object trajectories would be of the same length and time-stamps of their corresponding elements would be equal. The distance between two trajectories satisfying these conditions could be computed using Euclidean distance, simply by summing the distance over all elements with equal time-stamps. In the real world, on the other hand, non-overlapping observation intervals, time shifts and different sampling rates are common. Although various trajectory comparison functions have been proposed to cope with these difficulties, this topic is still an ongoing research area.

Most trajectory comparison functions stem from four basic algorithms: (1) Euclidean distance, (2) Longest Common Subsequence (LCSS), (3) Dynamic Time Warping (DTW), and (4) Edit distance. We classify these algorithms into two groups with respect to penalties added per pair-wise element comparisons: real penalty functions and quantized penalty functions. Real penalty functions measure the distance in terms of the Euclidean distance between observations while quantized penalty functions increment the distance by values 0 or 1 at each step depending on spatial proximity of the compared observations. In the following subsections we explain crucial trajectory comparison functions briefly and provide the reasoning behind this classification. For a detailed discussion on characteristics of these algorithms, please refer to [7].

Significance of our privacy preserving trajectory comparison protocol is due to the fact that it is applicable to all comparison functions explained below. Furthermore, the protocol does not trade accuracy against privacy unlike previous work.

4.1   Comparison Functions with Real Penalty

Euclidean distance, Edit distance with Real Penalty (ERP) and DTW are the comparison functions with real penalty. Euclidean distance is a naive method based on comparing the corresponding observations of trajectories with the same length. The algorithm terminates in $O(n)$ time, returning the sum of real penalties. The Euclidean distance function is sensitive to time shifts and noise but the output is a metric value.

ERP [6] measures the minimum cost of transforming the compared trajectory to the source trajectory using insertion, deletion and replacement operations. The cost of each operation is calculated using real spatial distance values. The cost of replacing observation i with observation j is $dist(i, j)$, where $dist$ is the Euclidean distance. However in case of insertion (or deletion), added cost is the distance between the inserted (or deleted) observation and the constant observation value g, defined by the user. ERP compares all pairs of elements in the trajectories, returning a metric value in $O(n^2)$ time. The algorithm is resistant to time shifts but not to noise.

DTW was initially proposed for approximate sequence matching in speech recognition but is generalized to similarity search in time series by the authors of [17]. The algorithm is very similar to Edit distance but instead of insertions and deletions, stutters are used. The ith stutter on x dimension, denoted as $stutter_i(x)$, repeats the ith element and shifts following elements to the right. Computation cost is $O(n^2)$ as expected and the resultant distance value is non-metric. Allowing repetitions strengthens the algorithm against time shifts but does not help with noise.

4.2   Comparison Functions with Quantized Penalty

Trajectory comparison functions with quantized penalty are LCSS [16] and Edit distance on Real Sequence (EDR) [7]. Both algorithms try to match all pairs of elements in the compared trajectories and therefore have a computation cost of $O(n^2)$. A pair of observations is considered a match if they are close to each other in space by less than a threshold, ϵ. LCSS returns the length of the longest matched sequence of observations while EDR returns the minimum number of insertion, deletion or replacement operations required to transform one trajectory to the other. Although these algorithms are resistant to time shifts and noise, distance values are not metric.

5   Communication Phase

As explained before, the protocol for privacy preserving comparison of trajectories consists of two phases: communication phase and computation phase. In the communication phase, data holders exchange data among themselves and the third party (TP), who will carry out the computation phase and publish the clustering results.

Prior to the communication phase we assume that every involved party, including the third party, has already generated pair-wise keys. These keys are used as seeds to pseudo-random number generators which disguise the exchanged messages. The Diffie-Hellman key exchange protocol is perfectly suitable for key generation [8].

The dissimilarity matrix is an object-by-object structure. In case of spatio-temporal data, an entry $D[i][j]$ of the dissimilarity matrix D is the distance between trajectories of objects i and j calculated using any comparison function. In Section 6, we show that our privacy preserving comparison protocol is suitable for all comparison functions explained in Section 4. If trajectories of both i and j are held by the same site, this site can calculate their distance locally and send it to the third party. However, if trajectories of i and j are at separate sites, these sites should run the protocol explained below. Assuming K data holders, $C(K, 2)$ runs are required, one for each pair of data holders.

Algorithm 1. Trajectory comparison protocol at site DH_A

Require: Trajectory database DH_A

1: **for** $j = 0$ to $size(DH_A) - 1$ **do**
2:     Initialize rng_{AB} with key K_{AB}
3:     Initialize rng_{AT} with key K_{AT}
4:     **for** $m = 0$ to $length(DH_A[j]) - 1$ **do**
5:         $DH_A[j][m].x = rng_{AT} + DH_A[j][m].x \times -1^{rng_{AB}\%2}$
6:         $DH_A[j][m].y = rng_{AT} + DH_A[j][m].y \times -1^{rng_{AB}\%2}$
7:     **end for**
8: **end for**
9:

Suppose that two data holders, DH_A and DH_B, with $size(A)$ and $size(B)$ trajectories respectively, want to compare their data. Assume that the protocol starts with DH_A. For each trajectory T in DH_A's database, two pseudo-random number generators are initialized, rng_{AB} and rng_{AT}. The seed for rng_{AB} is the key shared with DH_B and the seed for rng_{AT} is the key shared with TP. Then, for each spatial dimension of T's elements (i.e. x and y), DH_A disguises its input as follows: if the pseudo-random number generated by rng_{AB} is odd, DH_A negates its input and increments it by the pseudo-random number generated by rng_{AT}. Finally, DH_A sends the disguised values to DH_B.

Upon receiving data from DH_A, DH_B initializes a matrix M of size $size(B) \times size(A)$, which will be DH_B's output. For each trajectory T in its database, DH_B initializes a pseudo-random number generator rng_{AB} with the key shared with DH_A and negates its inputs in a similar fashion. This time negation is done when the generated number is even. DH_B then starts filling values into M. An entry $M[i][j][m][n]$ of M is DH_A's j^{th} trajectory's n^{th} observation compared to DH_B's i^{th} trajectories m^{th} observation. DH_B simply adds its input to the input received from DH_A. At the end, M is sent to TP by DH_B.

TP subtracts the random numbers added by DH_A using a pseudo-random number generator, rng_{AT}, initialized with the key shared with DH_A. Now, absolute value of any entry $M[i][j][m][n]$ is $|DH_A[j][n] - DH_B[i][m]|$. These values are all that is needed by any comparison function to compute the distance between trajectories i and j.

Pseudo codes for the roles described above are given in Algorithms 1, 2 and 3. Discussion on the necessity of each pseudo-random number generator used in the protocol is provided in Section 7.

6   Computation/Aggregation Phase

The third party can compute pair-wise trajectory distances for data holder sites A and B, once the comparison matrix M is built through the protocol in Section 5. If the comparison function measures distances using real penalty, then $M[i][j][m][n]$ is the cost for A's j_{th} trajectory's n_{th} observation with respect to

---

Algorithm 2. Trajectory comparison protocol at site DH_B

Require: Trajectory databases DH_A and DH_B

  1: **for** $i = 0$ to $size(DH_B) - 1$ **do**
  2:    **for** $j = 0$ to $size(DH_A) - 1$ **do**
  3:      **for** $n = 0$ to $length(DH_B[i]) - 1$ **do**
  4:       Initialize rng_{AB} with the key K_{AB}
  5:       **for** $m = 0$ to $length(DH_A[j]) - 1$ **do**
  6:         $M[i][j][n][m].x \mathrel{+}= DH_B[i][n].x \times -1^{(rng_{AB}+1)\%2}$
  7:         $M[i][j][n][m].y \mathrel{+}= DH_B[i][n].y \times -1^{(rng_{AB}+1)\%2}$
  8:      **end for**
  9:    **end for**
10:   **end for**
11: **end for**
12: Send M to TP

Algorithm 3. Trajectory comparison protocol at site TP

Require: Trajectory databases DH_A and DH_B

  1: **for** $i = 0$ to $size(DH_B) - 1$ **do**
  2:    **for** $j = 0$ to $size(DH_A) - 1$ **do**
  3:      **for** $n = 0$ to $length(DH_B[i]) - 1$ **do**
  4:       Initialize rng_{AT} with the key K_{AT}
  5:       **for** $m = 0$ to $length(DH_A[j]) - 1$ **do**
  6:         $M[i][j][n][m].x = |M[i][j][n][m].x - rng_{AT}|$
  7:         $M[i][j][n][m].y = |M[i][j][n][m].y - rng_{AT}|$
  8:      **end for**
  9:    **end for**
10:   **end for**
11: **end for**

B's i_{th} trajectory's m_{th} observation. Otherwise, if a quantized penalty comparison function is to be employed, TP simply checks whether $M[i][j][m][n] < \epsilon$ to match these two observations.

What remains is performing comparisons of the form $M[i][j]$, where both i and j are trajectories of the same data holder site. In such cases, another privacy preserving protocol is not required to compute these values, since conveying local dissimilarity matrices to TP does not leak any private information, proven in [13].

In order to build the dissimilarity matrix, TP must ensure that every data holder site has sent its local dissimilarity matrix and run the pair-wise comparison protocol with every other data holder. Algorithm 4 is the pseudo-code for constructing local dissimilarity matrices where $distance$ denotes the comparison function.

Algorithm 4. Local dissimilarity matrix construction

Require: Trajectory database DH
1: **for** $m = 0$ to $size(DH) - 1$ **do**
2:    **for** $n = 0$ to $m - 1$ **do**
3:      $D[m][n] = distance(DH[m], DH[n])$
4:    **end for**
5: **end for**

After gathering comparison results for all pairs of trajectories, TP normalizes the values in the dissimilarity matrix. These normalized distances are the only required input for most clustering algorithms, such as k-medoids, hierarchical and density based clustering algorithms. Another key observation here is that using our protocol, TP may use any clustering algorithm depending on requirements of the data holders.

At the end of the clustering process, the third party sends the clustering results to the data holders. The results are in the form of lists of objects identifiers, since publishing the dissimilarity matrix would cause private information leakage. The third party can also publish clustering quality parameters, if requested by the data holders.

7   Complexity and Privacy Analysis

In this section, we analyze the communication and computation costs of the pair-wise comparison protocol and local dissimilarity matrix construction. An analysis of the privacy offered by the protocol follows.

Every data holder has to send its local dissimilarity matrix to the third party. Computation cost of constructing the matrix is $O(n^2 \times distance)$ where n is the number of trajectories and $distance$ denotes the complexity of the comparison function. For Euclidean, the cost becomes $O(n^2 \times p)$ and for the other comparison functions it is $O(n^2 \times p^2)$ where p is the maximum number of observations in a trajectory.

The initiator of the comparison protocol, DH_A in Section 5, has a computation cost of $O(n \times p)$. The follower, DH_B, on the other hand makes $O(n \times m \times p^2)$ computations where m is the number of trajectories at site DH_B. Communication costs are parallel to computation costs since every party sends the result of the computation without any further operation.

There is an apparent imbalance in the computation and communication costs of the follower and initiator parties. TP can easily solve this problem by arranging the sequence that pair-wise comparison protocols are carried out such that every party will be the initiator at least $\lfloor (K-1)/2 \rfloor$ times in a setting of K data holders.

Sharing dissimilarity matrices does not leak any private information according to [13], as long as the private data is kept secret. The proof of the theorem relies on the fact that given the distance between two data points, there are infinitely many pairs of points that are equally distant. Since we assume that involved parties do not collude with each other and honestly follow the protocol, TP can not collude with a data holder site to infer private information of another data holder. Therefore sharing local dissimilarity matrices does not harm privacy unless the comparison protocol introduces inference channels that may leak private information.

In the comparison protocol, the message sent by the initiator is a matrix containing values of the form $(n+r)$ or $(-n+r)$ where n is initiator's input and r is a random number. In either case, these values are completely random to the follower. On the other hand, follower sends TP a matrix of values of the form $(n-m+r)$ or $(m-n+r)$. Although TP knows r, $(n-m)$ or $(m-n)$ does not help inferring neither n nor m, since there are infinitely many pairs (m,n) whose distance is $|m-n|$.

Purpose of the pseudo-random number generator shared between the initiator and the follower is preventing TP from inferring whose input is larger. Suppose that always the follower subtracts its input from the initiator's input. If $m > n$, $(n+r-m-r) = (n-m)$ would be negative, pointing out that follower's input is greater. The shared pseudo-random number generator garbles the negation sequence and prevents such inferences.

One possible attack against our comparison protocol could be statistical analysis. Notice that observations of every trajectory in initiator's database with the same index is disguised using the same random number. This is due to the fact that the pseudo-random number generator is re-initialized at each step. Given enough statistics on the data and assuming that the databases are large enough to contain many repetitions of spatial values, such an attack is realizable. But considering that the domain of spatial values is very large and such statistics is not publicly available, we regard these types of attacks as very unlikely to succeed.

8   Conclusions

In this chapter, we proposed a protocol for privacy preserving comparison of trajectories and its application to clustering of horizontally partitioned spatio-temporal data. The main advantage of our protocol is its applicability to most trajectory comparison functions and different clustering methods such as hierarchical clustering. The data holder sites can decide the clustering algorithm of their choice and receive clustering quality parameters together with the results. Only a small share of existing privacy preserving clustering algorithms can handle horizontally partitioned data and these algorithms do not specifically address spatio-temporal attributes.

We also provided complexity and privacy analysis of our protocol and observed that communication and computation costs are parallel to the computation costs for clustering local data. Privacy analysis shows that an attack using statistics of spatial components is possible but very unlikely to succeed. A proof-of-concept implementation of the clustering algorithm is available at [1]. We used real spatio-temporal datasets from the R-Tree Portal [2] for debugging and verifying the software.

References

1. ppstclusteringonhp.zip (3510k) (2006),
   http://students.sabanciuniv.edu/~inanali/ppSTClusteringOnHP.zip
2. The r-tree portal (2006),
   http://isl.cs.unipi.gr/db/projects/rtreeportal/trajectories.html
3. Agrawal, R., Srikant, R.: Privacy-preserving data mining. SIGMOD Rec. 29(2), 439–450 (2000)
4. Beresford, A.R., Stajano, F.: Mix zones: user privacy in location-aware services. In: Proceedings of the Second IEEE Annual Conference on Pervasive Computing and Communications Workshops, pp. 127–131 (2004)
5. Beresford, A.R.: Location Privacy in Ubiquitous Computing. PhD thesis, University of Cambridge (2004)
6. Chen, L., Ng, R.: On the marriage of lp-norms and edit distance. In: VLDB 2004: Proceedings of the Thirtieth International Conference on Very Large Data Bases, VLDB Endowment, pp. 792–803 (2004)
7. Chen, L., Özsu, M.T., Oria, V.: Robust and fast similarity search for moving object trajectories. In: SIGMOD 2005: Proceedings of the 2005 ACM SIGMOD International Conference on Management of Data, pp. 491–502. ACM, New York (2005)
8. Diffie, W., Hellman, M.E.: New directions in cryptography. IEEE Transactions on Information Theory IT-22(6), 644–654 (1976)
9. Hoh, B., Gruteser, M.: Protecting location privacy through path confusion. In: SECURECOMM 2005: Proceedings of the First International Conference on Security and Privacy for Emerging Areas in Communications Networks, Washington, DC, USA, pp. 194–205. IEEE Computer Society, Los Alamitos (2005)
10. Inan, A., Saygyn, Y., Savas, E., Hintoglu, A.A., Levi, A.: Privacy preserving clustering on horizontally partitioned data. In: ICDEW 2006: Proceedings of the 22nd International Conference on Data Engineering Workshops, Washington, DC, USA, p. 95. IEEE Computer Society, Los Alamitos (2006)
11. Jha, S., Kruger, L., Mcdaniel, P.: Privacy preserving clustering. In: Proceedings of the 10th European Symposium on Research in Computer Security, pp. 397–417 (2005)
12. Merugu, S., Ghosh, J.: Privacy-preserving distributed clustering using generative models. In: ICDM 2003: Proceedings of the Third IEEE International Conference on Data Mining, Washington, DC, USA, p. 211. IEEE Computer Society, Los Alamitos (2003)
13. Oliveira, S.R.M., Agropecuária, E.I., Tosello, A., Geraldo, B., Brasil, C.S.: Privacy-preserving clustering by object similarity-based representation and dimensionality reduction transformation. In: Proc. of the Workshop on Privacy and Security Aspects of Data Mining (PSADM 2004), in Conjunction with the Fourth IEEE International Conference on Data Mining (ICDM 2004), pp. 21–30 (2004)
14. Saygin, Y., Verykios, V.S., Clifton, C.: Using unknowns to prevent discovery of association rules. SIGMOD Rec. 30(4), 45–54 (2001)

15. Vaidya, J., Clifton, C.: Privacy-preserving k-means clustering over vertically partitioned data. In: KDD 2003: Proceedings of the Ninth ACM SIGKDD International Conference on Knowledge Discovery and Data Mining, pp. 206–215. ACM, New York (2003)
16. Vlachos, M., Gunopoulos, D., Kollios, G.: Discovering similar multidimensional trajectories. In: ICDE 2002: Proceedings of the 18th International Conference on Data Engineering, Washington, DC, USA, p. 673. IEEE Computer Society, Los Alamitos (2002)
17. Yi, B.-K., Jagadish, H.V., Faloutsos, C.: Efficient retrieval of similar time sequences under time warping. In: ICDE 1998: Proceedings of the Fourteenth International Conference on Data Engineering, Washington, DC, USA, pp. 201–208. IEEE Computer Society, Los Alamitos (1998)

Nemoz — A Distributed Framework for Collaborative Media Organization

Katharina Morik

Technical University Dortmund, Computer Science LS8
44221 Dortmund, Germany
morik@ls8.cs.tu-dortmund.de

Abstract. Multimedia applications have received quite some interest. Embedding them into a framework of ubiquitous computing and peer-to-peer Web 2.0 applications raises research questions of resource-awareness which are not that demanding within a server-based framework. In this chapter, we present Nemoz, a collaborative music organizer based on distributed data and multimedia mining techniques. We introduce the Nemoz platform before focusing on the steps of intelligent collaborative structuring of multimedia collections, namely, feature extraction and distributed data mining. We summarize the characteristics of knowledge discovery in ubiquitous computing that have been handled within the Nemoz project.

1 Introduction

The Nemoz system has already been mentioned to illustrate ubiquitous intelligent media organization (Chap. 1, Sec. 6.2) and peer-to-peer based Web 2.0 applications (Chap. 4, Sec. 2.3). Here, a more detailed presentation of Nemoz exemplifies the large variety of problems which have to be solved in order to develop knowledge discovery in ubiquitous computing. Some of the problems are due to the nature of audio data or, more generally, of unstructured data. In particular, the problem of *feature extraction* is demanding. Although this problem has to be handled also in server-based systems, the distributed setting as well as the resource restrictions regarding runtime and communication hampers its solution considerably. Some of the problems are due to the heterogeneity of users as we face it in social networks. Our approach of *aspect-based tagging* is applicable also in centralized systems, but again, the setting of ubiquitous computing challenges us further because of the resource restrictions and the unreliability of communication. Some of the problems like, e.g., data indexing and system design, are part of every system development. It might be of interest to look at the methods regarding audio data, social networks, and system design in concert and embedded into the setting of knowledge discovery in ubiquitous computing.

In this chapter we present Nemoz, a collaborative music organizer based on distributed data and multimedia mining techniques. The organization of music is a an ideal test case for distributed multimedia systems. First, the management of

M. May and L. Saitta (Eds.): Ubiquitous Knowledge Discovery, LNAI 6202, pp. 199–215, 2010.

audio data is a very hard task because the corresponding semantic descriptions depend on social and personal factors. The heterogeneity of users has to be taken into account. Second, audio collections are usually inherently distributed. Hence, the ubiquity of the data is to be handled. Third, audio data is stored and managed on a large variety of different devices with very different capabilities concerning network connection and computational power. The availability of data is varying. All this has a strong influence on the methods that can be used.

Currently, large amounts of multimedia content are stored at personal computers, MP3 devices or mobile phones, and other devices. The organization of these collections is cumbersome. There is a trade-off concerning personalization and effort. On the one hand, designing individual structures takes time, but personalizes the collection's organization perfectly. On the other hand, accepting a given structure like that offered by iTunes is easy, but does not reflect the individual view of the user. While users might start to design their own structures for their collection, they are not so enthusiastic about maintaining the structures and tagging every new item. The key idea is now to exploit the partial annotations of other users for *personalized, automatic tagging* of the own (partially annotated) collection.

The chapter is structured as follows: Section 2 introduces the Nemoz platform. Section 3 distinguishes the problem of feature extraction in the setting of ubiquitous knowledge discovery. In Section 4, we analyze how aspect-enriched tagging structures can be exploited by distributed data mining methods. In Section 8, we summarize where the characteristics of knowledge discovery in ubiquitous computing showed up in the Nemoz project[1].

2   Nemoz - Networked Media Organizer

As has been pointed out in the introduction, (Chap. 1, Sec. 4), ubiquitous knowledge discovery is always integrated. Hence, an overview of the system which exploits knowledge discovery is given in the following. Each peer node has to install Nemoz in order to structure and use its own collection and participate in the collaborative media organization.

2.1   The Functionality

As a media organizer, Nemoz implements the basic functions of media systems:

- download and import of songs,
- playing music,
- browsing through a collection,
- retrieving music from a collection based on given meta data, and
- creating play lists.

---

[1] The Nemoz project started as a student project and was continued by Oliver Flasch, Andreas Kaspari, Michael Wurst, and the author.

Each user may create arbitrary, personal hierarchies of tags (i.e., taxonomies) in order to organize her music, for instance with respect to mood and situations, or to genres, etc. Some of these structures may overlap, e.g., the "blues" genre may cover songs which are also covered by a personal concept "melancholic" of a structure describing moods. Figures 1 and 2 show two views of such personal structures. Nemoz supports the users in structuring their media objects while not forcing them to use a given global set of concepts or annotations.

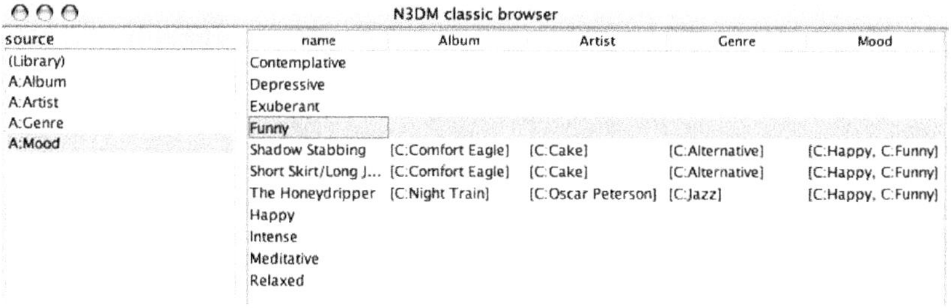

source	name	Album	Artist	Genre	Mood
(Library)	Contemplative				
A.Album	Depressive				
A.Artist	Exuberant				
A:Genre	Funny				
A:Mood	Shadow Stabbing	[C:Comfort Eagle]	[C.Cake]	[C.Alternative]	[C:Happy, C.Funny]
	Short Skirt/Long J...	[C.Comfort Eagle]	[C.Cake]	[C.Alternative]	[C:Happy, C:Funny]
	The Honeydripper	[C.Night Train]	[C:Oscar Peterson]	[C:Jazz]	[C:Happy, C.Funny]
	Happy				
	Intense				
	Meditative				
	Relaxed				

(N3DM classic browser)

Fig. 1. Aspects in our formalism form trees. The Nemoz taxonomy browser shows here the "mood" taxonomy and the extension of the mood "funny".

Functionality based on ubiquitous knowledge discovery are:

- Tag nodes can intensionally be defined by a (learned) model.
- Tags can be automatically assigned to new items.
- Users can search for similar aspects and tags in the network.
- Users can search for music similar to a selected song or group of songs concerning a (personal) aspect.
- Users can view a collection (of another user) conforming to a selected (own) tag structure ("goggle").
- Tag structures can be automatically enhanced through the tags of other users (see Section 4).

Technically, tags can be interpreted as concepts and aspects can be interpreted as individual taxonomies (thus as the hierarchical part of an ontology). An extended function is stored at each node of the taxonomy, which decides whether a new song belongs to this node, or not. Since tags are chosen arbitrarily by the users, we cannot rely on a predefined scheme. Rather, the classifier is a learned model, where the learning task has been performed by one of the methods supplied by the machine learning environment RapidMiner [10][2]. Applying the models learned from a collection annotated with respect to $aspect_1$ to another collection, structured according to another $aspect_2$, allows to goggle at a collection "through the own glasses".

---

[2] RapidMiner was formerly known as YALE. It is available at http://rapid-i.com/

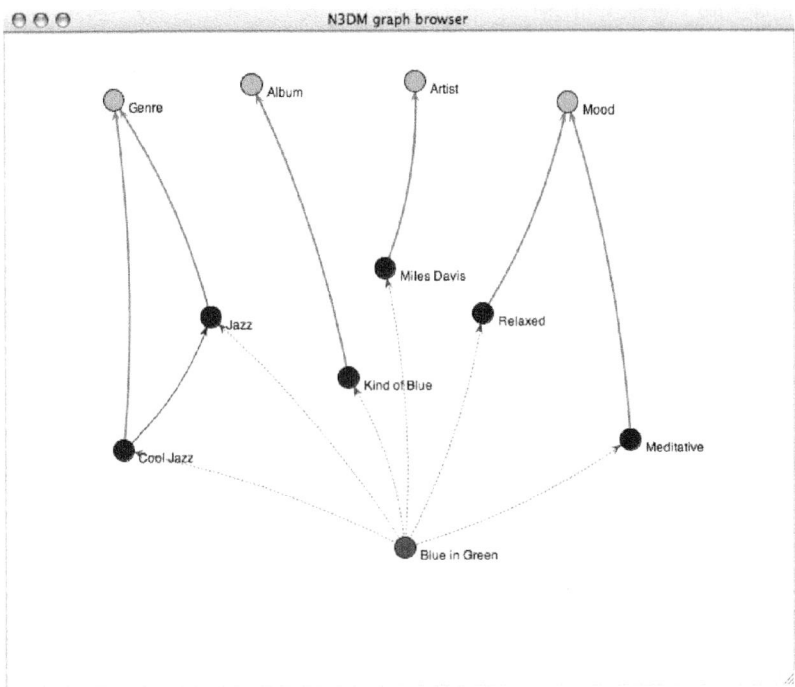

Fig. 2. The Nemoz graph browser is especially useful when examining items that have been structured under more than one aspect

2.2   The Data Model and Services

The data model covers not only standard music meta-data (performer, composer, album, year, duration of the song, genre, and comment) and a reference to the location of the song, but also features which are extracted from the raw sample data (cf. Section 3). Confronted with music data, machine learning encounters a new challenge of scalability. Music databases store millions of records and each item contains up to several million values. Hence, most processing has to be performed on identifiers or indices, the music records must neither be moved around nor be copied. The basic entities of the data model are *users*, *items* (songs), *categories*, and *aspects*. Instead of storing these entities directly, we distinguish between abstract, opaque entity identifiers and entity representations. This distinction is motivated by the "Representational State Transfer" [3] paradigm of the World Wide Web, to which our formalism adheres to. The *Descriptor Service* delivers unique identifiers for an object and manages the link to the full description. In particular, the full set of features of an object is only retrieved if necessary. This is important for the reduction of network communication. The service handles locking in case of modifications of the same object by several threads as well as caching of full descriptions in case of continued processing requiring the full description. The Descriptor Service interacts with the so-called *Cloakroom Service*, which manages the objects storage and allocates free storage.

Representing all information consistently in the Web 2.0 *everything is a link* manner facilitates the implementation in a distributed environment. Items and tags are represented by category identifiers. Links between items and tags are the basic tagging. Our concept of a category extends the Web 2.0 tagging model by explicitly allowing "categories of categories", thereby enabling the representation of hierarchical structures akin to description logics [1]. A partial order link between categories c and c' need not be based on the extensions, but is valid even if extensionally there is not yet any difference. Consider, for instance, a user whose music library contains little jazz, all by Miles Davis. Our formalism would not force a user to accept the rather nonsensical identification of jazz and Miles Davis implied by the identity of the extension sets. If this identification actually reflects the user's opinion, she is still free to declare it explicitly.

The formalism allows the user to organize categories further by grouping them into aspects through a link between categories and aspects. This is handled by the *Taxonomy Service*. Typical examples for aspects from the music domain are "genre", "mood", "artist" and "tempo". The usefulness of aspects has several facets. First, hierarchical category structures tend to become unmanageable when growing in size. Aspects enable the user to create complex structures to organize her items and simultaneously maintain clarity. Consider a user, who uses del.icio.us to organize her hyperlinks. With a great number of tags, retrieving one such link becomes more and more complicated. Grouping tags/categories into aspects eases this task considerably. Second, aspects can be used for filtering large category structures. Filtering means restricting the visible fraction of these structures to a specific topic. A limited variant of this notion is implemented in the iTunes media organizer, where the user can select a genre or an artist she wants to browse. Nemoz enables the user to browse her items by arbitrary aspects. Third, aspects implicitly define a similarity measure on items that can be used to realize aspect-based clustering and visualization.

In our formalism, entities are ownerless, only links are owned by users. Each link must have at least one owner. It may have multiple owners, if it has been added independently by multiple users. Item ownership is not a first class concept in our formalism. Our user concept comprises human users as well as *intelligent agents*. An agent acts on behalf of a human user, but has an identity of its own. For example, the intelligent operations (i.e., clustering and classification) have been modeled using such agents. Each time, an intelligent operation is triggered by a user, an agent user is created that performs the operation and adds the resulting links to the knowledge base. Our design gives the user control over the effects of these operations by clearly distinguishing between automatically generated and manually entered knowledge. An automatically generated link may be promoted to a user-approved link by changing the link ownership from an agent to its client user. The effects of an intelligent operation may be canceled by deleting the responsible agent. By keeping automatically generated knowledge in an ephemeral state until it has been approved by the user, we hope to tame the effects of a possibly poor performing intelligent operation.

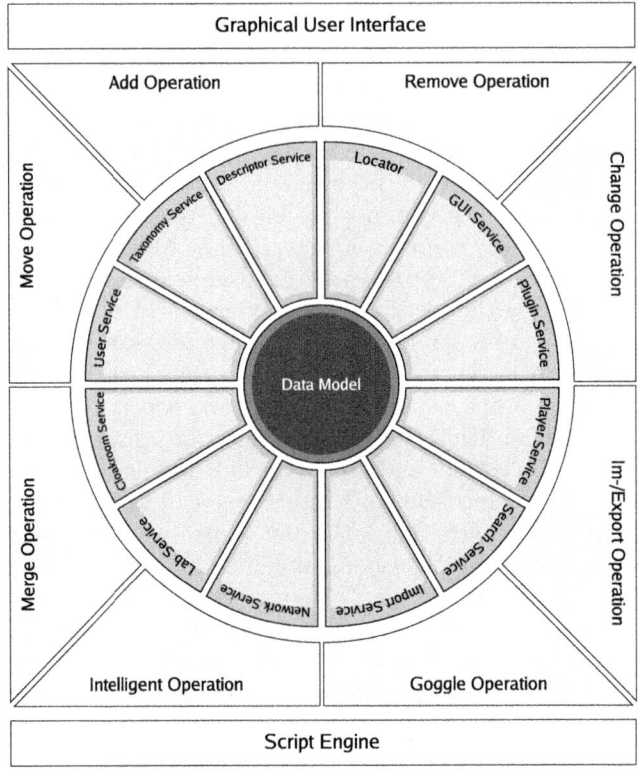

Fig. 3. The Nemoz architecture shows the services grouped around the data model. All services access the data model and provide the operations with a coherent abstract interface. While Nemoz is written in Java, the Script Engine is written in Scheme and allows the integration and interpretation of new scripts, thus supporting rapid prototyping.

Learning taxonomies is performed using RapidMiner (see Section 4). The *Lab Service* couples Nemoz with the learning engine RapidMiner providing it with sets of examples and retrieving the learned models.

2.3   The Peer-to-Peer Network

The *Search Service* and the *Import Service* are designed for ad hoc networks. The *Network Service* supports the communication among (W)LAN nodes via TCP for direct peer-to-peer interaction and UDP for broadcasting. Since Jtella did not work well, the Network Service has been implemented using JavaSockets. It could as well have been implemented using JXTA. The logon/logout of users, sending a ping message around, and starting threads for new connections is the basic functionality. Browsing through other peers' collections is implemented in a node-wise manner, where first only reduced descriptions of tags are transferred. Only if a user clicks on a taxonomy node, its full description and the restricted description of the child nodes is sent. The privacy attribute of nodes prevents private data from being communicated to another peer.

The Search Service works asynchronously. A query is a uniquely identified quantifier-free logical formula without negation. The query is broadcasted and answers are accepted within a specified time interval. Answers to one query are bundled together. The search result is a set of three lists (for descriptors, aspects (i.e., taxonomies), users), each consisting of pairs $< result, relevance >$. Since the collections are indexed by lexical trees (Patricia tries [6]), substrings of the string indices can be exploited for substring search. A metrical index eases similarity search. It can directly be used by the k-NearestNeighbor-method.

Where the Search Service works on restricted descriptions of data, the Import Service actually retrieves the audio data, either as a small fraction from the middle of a song, or as the complete song.

3   Feature Extraction

While meta-data like "artist", "year", "genre", and tags are regular nominal features, the audio data are value series. The shape of the curve defined by these values does not express the crucial aspect of similarity measures for musical objects. The solution to overcome these issues is to extract features from the audio signal which leads to a strong compression of the data set at hand. A large variety of approaches has been proposed for manually extracting features from audio data [5,14].

However, it turns out that optimal audio features strongly depend on the task at hand [12] and the current subset of items [11]. It is not very likely that a feature set delivering excellent performance on the separation of classical and popular music works well also for the separation of music structured according to occasions. This problem already arises for high-level structures like musical genres and is even aggravated due to the locality induced by personal structures. One possibility to cope with this problem is to learn an adapted set of features for each learning task separately [7,18]. These approaches achieve a high accuracy, but are computationally very demanding and not well suited for distributed settings.

If there would exist one complete set of features, from which each learning task selects its proper part, the feature problem could be reduced to feature selection. However, the number of possible feature extractions is so large – virtually infinite – that it would be intractable to enumerate it. Hence, for practical reasons we have chosen a compromise. Some feature sets are trained locally for a learning task on a local collection using the method described in [7]. Now, some of the learning tasks may turn out to be best solved using (almost) the same feature set. These tasks are considered similar. For each task, the usefulness of features from a basic feature set is determined. A bit vector of the basic features with value 1 indicating that a feature is useful, value 0 indicating that it is not, indexes learning tasks. Tasks with the same index retrieve the full, trained feature set. Hence, the efforts for training feature sets are restricted, but its use becomes available for a large range of learning tasks in a case-based manner [9]. Most frequently, only the index vector needs to be communicated in the network[3].

4   Aspect-Based Distributed Multimedia Mining

Based on user taggings, several data mining techniques can be applied. The aim of the techniques is to support users in organizing and navigating their media collections. In the following we will focus on two tasks: first, the question of how audio files can be automatically annotated given a set of tagged examples and second the question of how to structure a set of audio files exploiting taggings of others. For both tasks, distributed data mining methods are provided.

By recommending tags and structures to other users, views on the underlying space of objects are emerging. This approach naturally leads to a social filtering of such views. If someone creates a (partial) taxonomy found useful by many other users, it is often copied. If several taxonomies equally fit a query, a well-distributed taxonomy is recommended with higher probability. This pushes high quality taxonomies and allows to filter random or non-sense taxonomies.

4.1   Automatic Tagging

Often it is important to assign tags to audio files automatically. Users need to tag only a small amount of audio files manually, while the remainder of the audio files is tagged automatically. The learned classifiers are applicable not only to the set of music from which they have been learned, but also to other structures. Hence, tagging audio files automatically allows to browse ("goggle") music collections of other users using ones own terms.

For the automatic tagging of audio files, we use the aspect representation of taxonomies together with hierarchical classification. All concepts belonging to an aspect must represent a tree. We tag audio files by training a classifier for each node in a concept hierarchy. Then, for each audio file these classifiers are applied in a top down manner until a leaf node is reached. This procedure is applied for each aspect and each audio file. Thus each audio file receives one most specific tag per aspect, which is connected to several super concepts according to the concept relation.

Aspects define the entry points for hierarchical classification (each aspect is a tree with exactly one root node) and group concepts explicitly into a set of trees. Without aspects, the algorithm would face a huge number of binary classification problems (one for each tag) for which furthermore negative examples are not explicitly given.

---

[3] For more details on the collaborative use of features in distributed systems see [8].

In the peer-to-peer setting, we use tags of other users as features when train-ing classifiers. This approach is described in [16]. The idea is that even though tags of different users may not be identical, they can still be highly useful for classification. For example, background music is not identical to jazz, could how-ever together with an additional audio feature, allow to perform classification with higher accuracy. Therefore, nodes can query other nodes for tags, which are then used as features.

4.2   Unsupervised Tagging

If a user has not yet assigned tags, these can be inferred by clustering. Traditional feature based clustering does not produce satisfying results, because it is hard to infer labels for clusters, which is the basis for tagging.

In [17] we introduced the LACE method which combines tags of other users for tagging one's own music collection. The idea is to cover the items to be tagged with a set of concept structures obtained from other users. This allows to cluster even heterogeneous sets of items, because different subsets may be covered by different clusterings. Items that cannot be covered by clusterings obtained from other users are simply assigned to tags using classification, as described above. Figure 4 shows the main idea. LACE is not an instance of standard distributed clustering. Distributed clustering learns a global model integrating the various local ones [2]. However, this global consensus model would destroy the structure already created by the user and does not focus on the set S of not appropriately structured items.

The new learning task can be formally stated. Let X denote the set of all possible objects. A function $\varphi : S \to G$ is a function that maps objects $S \subseteq X$ to a (finite) set G of groups. We denote the domain of a function φ with D_φ. In cases where we have to deal with overlapping and hierarchical groups, we denote the set of groups as 2^G.

Given a set $S \subseteq X$, a set of input functions $I \subseteq \{\varphi_i : S_i \to G_i\}$, and a quality function

$$q : 2^\Phi \times 2^\Phi \times 2^S \to R \tag{1}$$

with R being partially ordered

localized alternative clustering ensembles delivers the output functions $O \subseteq \{\varphi_i | \varphi_i : S_i \to G_i\}$ so that $q(I, O, S)$ is maximized and for each $\varphi_i \in O$ it holds that $S \subseteq D_{\varphi_i}$.

Note that in contrast to cluster ensembles [13], the input clusterings can be defined on any subset S_i of X. Since for all $\varphi_i \in O$ it must hold that $S \subseteq D_{\varphi_i}$, all output clusterings must at least cover the items in S.

The *quality of an individual output function* is measured as

$$q^*(I, \varphi_i, S) = \sum_{x \in S} \max_{x' \in S_{ij}} sim(x, x') \text{ with } j = h_i(x) \tag{2}$$

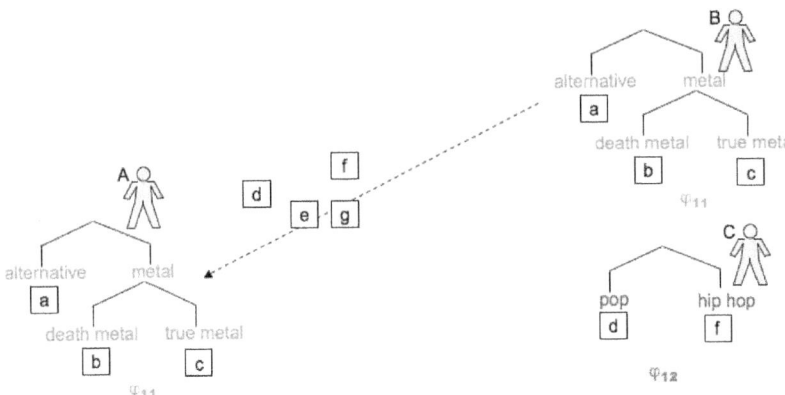

Fig. 4. The picture simplifies a state in LACE processing, where User A posts the demand to structure items a, b, c, d, e, f, g forming the set S. She has received a structure φ_{11} from user B covering the before untagged songs a, b, c. For simplicity, here the songs in the taxonomy φ_{11} of B are identical to some of A's unstructured ones. For the remaining items $d, e, f, g \in S$, a second posting will deliver the structure φ_{12} from user C covering items d, f. The structures φ_{11} and φ_{12} will be combined. If the remaining items e, g cannot be covered by another user's taxonomy, they will be sorted into the most similar node of user A's new combined taxonomy.

where sim is a similarity function $sim : X \times X \rightarrow [0, 1]$ and h_i assigns each example to the corresponding function in the bag of clusters $h_i : S \rightarrow \{1, \ldots, m\}$ with

$$h_i(x) = j \Leftrightarrow x \in S'_{ij} \tag{3}$$

The *quality of a set of output functions* now becomes

$$q(I, O, S) = \sum_{\varphi_i \in O} q^*(I, \varphi_i, S) \tag{4}$$

Besides this quality function, we want to cover the set S with a bag of clusterings that contains as few clusterings as possible.

Given a set I of functions
a Bag of clusterings is a function

$$\varphi_i(x) = \begin{cases} \varphi'_{i1}(x), & \text{if } x \in S'_{i1} \\ \vdots & \vdots \\ \varphi'_{ij}(x), & \text{if } x \in S'_{ij} \\ \vdots & \vdots \\ \varphi'_{im}(x), & \text{if } x \in S'_{im} \end{cases} \tag{5}$$

where each φ'_{ij} is an extension of a $\varphi_{ij} \in I$ and $\{S'_{i1}, \ldots, S'_{im}\}$ is a partitioning of S.

Extended function: Given a function $\varphi_i : S_i \rightarrow G_i$, the function $\varphi_i' : S_i' \rightarrow G_i$ is the *extended function* for φ_i, if $S_i \subset S_i'$ and $\forall x \in S_i : \varphi_i(x) = \varphi_i'(x)$.

The main task is to cover S by a bag of clusterings φ. The basic idea of this approach is to employ a sequential covering strategy. In a first step, we search for a function φ_i in I that best fits the set of query objects S. For all objects not sufficiently covered by φ_i, we search for another function in I that fits the remaining points. This process continues until either all objects are sufficiently covered, a maximal number of steps is reached, or there are no input functions left that could cover the remaining objects. All data points that could not be covered are assigned to the input function φ_j containing the object which is closest to the one to be covered. Alternative clusterings are produced by performing this procedure several times, such that each input function is used at most once.

When is a data point sufficiently covered by an input function so that it can be removed from the query set S? We define a threshold based criterion for this purpose. A function φ *sufficiently covers* a object $x \in S$ (written as $x \sqsubset_\alpha \varphi$), iff

$$x \sqsubset_\alpha \varphi :\Leftrightarrow \max_{x' \in Z_\varphi} sim(x, x') > \alpha \tag{6}$$

This threshold allows to balance the quality of the resulting clustering and the number of input clusters. A small value of α allows a single input function to cover many objects in S. This, on average, reduces the number of input functions needed to cover the whole query set. However, it may also reduce the quality of the result, because the algorithm then covers many objects, which could have been covered better using an additional input function. A large value of α combines many input functions. However, this might yield groups with too few instances.

Turning it the other way around: when do we consider an input function to fit the items in S well? The situation we are facing is similar to that in information retrieval. The target concept S – the ideal response – is approximated by φ delivering a set of items – the retrieval result. If all members of the target concept are covered, the retrieval result has the highest recall. If no items in the retrieval result are not members of S, it has the highest precision. Hence, we have tailored precision and recall to characterize how well φ covers S (see [17]).

Deciding whether φ_i fits S or whether an object $x \in S$ is sufficiently covered requires to compute the similarity between an object and a cluster. If the cluster is represented by all of its objects ($Z_{\varphi_i} = S_i$, as usual in single-link agglomerative clustering), this central step becomes inefficient. If the cluster is represented by exactly one point ($|Z_{\varphi_i}| = 1$, a centroid in k-means clustering), the similarity calculation is very efficient, but the centroid is not necessarily a member of S_i. Sets of objects with irregular shape, for instance, cannot be captured adequately. Hence, we adopt the representation of "well scattered points" Z_{φ_i} as representation of φ_i [4], where $1 < |Z_{\varphi_i}| < |S_i|$. This combines an increase in efficiency, as achieved by using a single centroid, with the ability to accurately represent data sets of irregular shape, as achieved by using all objects. In general, it is a non-trivial task to find such a set of well scattered points. However, in the context of our approach, the situation is easier, because groups are already given.

We can therefore draw a stratified sample of objects in S_i obtaining a set of well scattered points efficiently.

As each function is represented by a fixed number of representative points, the number of similarity calculations performed by the algorithm is linear in the number of query objects and in the number of input functions, thus $O(|I||S||Z_{\varphi_i}|)$. The same holds for the memory requirements.

$$
\begin{aligned}
&O = \emptyset \\
&J = I \\
&\textbf{while } (|O| < max_{alt}) \textbf{ do} \\
&\quad S_u = S \\
&\quad B = \emptyset \\
&\quad step = 0 \\
&\quad \textbf{while } ((S_u \neq \emptyset) \wedge (step < max_{steps})) \textbf{ do} \\
&\qquad \varphi_i = \arg\max_{\varphi \in J} q_f^*(Z_\varphi, S_u) \\
&\qquad S_u = S_u \setminus \{x \in S_u | x \sqsubseteq_\alpha \varphi_i\} \\
&\qquad B = B \cup \{\varphi_i\} \\
&\qquad step = step + 1 \\
&\quad \textbf{end while} \\
&\quad O = O \cup \{bag(B, S)\} \\
&\textbf{end while}
\end{aligned}
$$

Fig. 5. The sequential covering algorithm finds bags of clusterings in a greedy manner. max_{alt} denotes the maximum number of alternatives in the output, max_{steps} denotes the maximum number of steps that are performed during sequential covering. The function bag constructs a bag of clusterings by assigning each object $x \in S$ to the function $\varphi_i \in B$ that contains the object most similar to x.

The algorithm described so far can only combine complete input clusterings and does not yet handle hierarchies. Extending it to the combination of partial hierarchical functions, it becomes LACE. A hierarchical function maps objects to a hierarchy of groups.

The set G_i of groups associated with a function φ_i builds a *group hierarchy*, iff there is a relation $<$ such that $(g < g') :\Leftrightarrow (\forall x \in S_i : g' \in \varphi_i(x) \Rightarrow g \in \varphi_i(x))$ and $(G_i, <)$ is a tree. The function φ_i is then called a *hierarchical function*.

We formalize this notion by defining a hierarchy on functions, which extends the set of input functions such that it contains all partial functions, as well.

Two hierarchical functions φ_i and φ_j, are in *direct sub function relation* $\varphi_i \prec \varphi_j$, iff

$$G_i \subset G_j$$

$$\forall x \in S_i : \varphi_i(x) = \varphi_j(x) \cap G_i$$

$$\neg \exists \varphi_i' : G_i \subset G_i' \subset G_j$$

Let the set I^* be the set of all functions which can be achieved following the direct sub function relation starting from I.

$$I^* = \{\varphi_i | \exists \varphi_j \in I : \varphi_i \prec^* \varphi_j\}$$

where $\prec^*$ is the transitive hull of $\prec$.

While it would be possible to apply the same algorithm as above to the extended set of input functions I^*, this would be rather inefficient, because the size of I^* can be considerably larger than the one of the original set of input functions I. Calculating the precision and recall of an input function with respect to items in S should now exploit the function hierarchy and avoid multiple similarity computations. Each function $\varphi_i \in I^*$ is again associated with a set of representative objects Z_{φ_i}. We additionally assume the standard taxonomy semantics:

$$\varphi_i \prec \varphi_j \Rightarrow Z_{\varphi_i} \subseteq Z_{\varphi_j}$$

Now, the precision can be calculated recursively in the following way:

$$prec(Z_{\varphi_i}, S) = \frac{|Z^*_{\varphi_i}|}{|Z_{\varphi_i}|} prec(Z^*_{\varphi_i}, S) + \sum_{\varphi_j \prec \varphi_i} \frac{|Z_{\varphi_j}|}{|Z_{\varphi_i}|} prec(Z_{\varphi_j}, S). \tag{7}$$

where $Z^*_{\varphi_i} = Z_{\varphi_i} \setminus \bigcup_{\varphi_j \prec \varphi_i} Z_{\varphi_j}$. For recall a similar function can be derived. Note, that neither the number of similarity calculations is greater than in the base version of the algorithm nor are the memory requirements increased.

Moreover, the bottom-up procedure also allows for pruning. We can optimistically estimate the best precision and recall, that can be achieved in a function hierarchy using all representative objects Z_e for which the precision is already known. The following holds:

$$prec(Z_{\varphi_i}, S) \le \frac{|Z_e| prec(Z_e, S) + |Z_{\varphi_i} \setminus Z_e|}{|Z_{\varphi_i}|}$$

with $Z_e \subset Z_{\varphi_i}$. An optimistic estimate for the recall is one.

If the optimistic f-measure estimate of the hierarchy's root node is worse than the current best score, this hierarchy does not need to be processed further. This is due to the optimistic score increasing with $|Z_{\varphi_i}|$ and $|Z_{\varphi_i}| > |Z_{\varphi_j}|$ for all subfunctions $\varphi_j \prec \varphi_i$. No subfunction of the root can be better than the current best score, if the score of the root is equal or worse than the current best score.

This conversion to hierarchical cluster models concludes our algorithm for local alternative cluster ensembles, LACE. For more details see [15].

The LACE algorithm is well suited for distributed scenarios as illustrated by Figure 4. We assume a set of nodes connected over an arbitrary communication network. Each node has one or several functions φ_i together with the sets S_i. If a node A has a set of objects S to be clustered, it queries the other nodes and these respond with a set of functions. The answers of the other nodes form the input functions I. A node B being queried uses its own functions φ_i as input and determines the best fitting φ_i for S and send this output back to A. The algorithm is the same for each node. Each node executes the algorithm independently of the other nodes.

We have applied three further optimizations. First, given a function hierarchy, each nodes returns exactly one optimal function in the hierarchy. This reduces the communication cost, without affecting the result, because any but the optimal function would not be chosen anyway.

Second, input functions returned by other nodes can be represented more efficiently by only containing the items in the query set, that are sufficiently covered by the corresponding function. Together with the f-measure value for the function, this information is sufficient for the querying node in order to perform the algorithm.

In many application areas, we can apply a third optimization technique. If objects are uniquely identified, such as audio files, films, web resources, etc. they can be represented by these IDs, only. Using this representation, a distributed version of our algorithm only needs to query other nodes using a set of IDs. This reduces the communication cost and makes matching even more efficient. Furthermore, such queries are already very well supported by current technology, such as peer-to-peer search engines. In a distributed scenario, network latency and communication cost must be taken into account. If objects are represented by IDs, both are restricted to an additional effort of $O(|S| + |I^*|)$. Thus, the algorithm is still linear in the number of query objects.

5   Discussion and Conclusion

In this chapter, we have exemplified some of the issues when developing a system for ubiquitous knowledge discovery. The music data as well as the hierarchical taggings are distributed among peers. Although the peers' collections differ, they do overlap. Hence, these are *vertically distributed data*. The resource limitations concerning memory and bandwidth are demanding. Since audio files are large, an important issue is their representation at several levels of granularity. This is handled by the Descriptor and the Cloakroom Service of Nemoz. The separation of objects and their descriptions is obeyed through-out the software at all levels.

- Most processes work just on unique identifiers. If such identifiers are already given, they are used. If not, a software service creates them and uses them for communication in the network. These identifiers then have to be mapped to local, internal identifiers.
- Where more information about an audio file is necessary, the peer locally retrieves the full file description. The description is composed of meta-data and extracted features. Even this full description compresses the object itself (the audio file) considerably.

- Only in a few cases, the full description of another peer needs to be transferred.
- Only on the user's demand, e.g., by selecting a taxonomy node in another peer's structure, the full audio file is transferred.

The work-share between peers needs to be carefully designed. Moreover, asynchronous processing must be supported. The Search Service of Nemoz asynchronously retrieves search results from peers, bundled together by unique identifiers. While this is straightforward, managing *feature extraction* is a key challenge to distributed multimedia management. The new technique of case-based feature transfer based on adaptive feature extraction balances performance gain of learning on the one hand and computation efforts on the other hand.

- The training of feature extractions is performed locally.
- Only index vectors are communicated. They identify classes of learning tasks for which a particular set of features is well suited .
- The feature extraction itself is, again, performed locally.

The new learning of *aspect-based tagging* generalizes the notion of a taxonomy for structuring multimedia collections. There is not one global taxonomy into which all items have to be classified, but each item can be covered by several taxonomies depending on the aspect. This is why we identify "taxonomy" and "aspect". A user is not identified with one taxonomy. The heterogeneity of users as well as occasions is taken into account by the notion of aspects. Hence, handling various aspects is the same, be they held by one or several peers. For learning in a distributed setting, however, communication restrictions need to be taken into account. LACE is the first algorithm, which distributedly learns personalized structures from heterogeneous taxonomies. LACE's similarity computation is linear in the number of query objects (i.e., songs) and the number of input functions (i.e., partial taxonomies) as is the memory requirement. Hierarchical structures are exploited to constrain the learning. The processes are asynchronous.

- A peer posting a request for structures that cover a set of items combines the answers locally at arrival time.
- All peers receiving a request perform a local fitness evaluation of given aspects and the set of items independently of each other.
- The relevance of a peer's structure for the request is rated locally at the side of the queried peer.

Organizing music in distributed systems is challenging. Users often have very different ideas of how to structure music. Using a global, fixed scheme is therefore not well-suited for this task. The success of Web 2.0 systems like last.fm stands evidence for this observation. Allowing users to organize their media using arbitrary classification schemes demands for new representation mechanisms and intelligent multimedia mining methods. Here, we move beyond the typical Web 2.0 systems in providing hierarchical structures for several aspects and learn new, personalized taxonomies from others. Since many current systems are not based on a client/server paradigm, but rather on decentralized adhoc networks, these multimedia mining methods must be distributed as well. The automatic support by machine learning techniques as well as the distributed setting with its collaborative approaches opens the floor for new ways of user collaboration and better services for users.

References

1. Baader, F., Calvanese, D., McGuinness, D., Nardi, D., Patel-Schneider, P.: The Description Logic Handbook. Cambridge University Press, Cambridge (2003)
2. Datta, S., Bhaduri, K., Giannella, C., Wolff, R., Kargupta, H.: Distributed data mining in peer-to-peer networks. IEEE Inteternet Computing, Special Issue on Distributed Data Mining (2005)
3. Fielding, R.T.: Architectural styles and the design of network-based software architectures. PhD thesis, University of California (2000)
4. Guha, S., Rastogi, R., Shim, K.: CURE: an efficient clustering algorithm for large databases. In: Proceedings of International Conference on Management of Data, pp. 73–84 (1998)
5. Guo, G., Li, S.Z.: Content-based audio classification and retrieval by support vector machines. IEEE Transaction on Neural Networks 14(1), 209–215 (2003)
6. Knuth, D.E.: The Art of Computer Programming. Addison-Wesley, Reading (1998)
7. Mierswa, I., Morik, K.: Automatic feature extraction for classifying audio data. Machine Learning Journal 58, 127–149 (2005)
8. Mierswa, I., Morik, K., Wurst, M.: Collaborative use of features in a distrbuted system for the organization of music collections. In: Shen, J., Shepherd, J., Cui, B., Ling, L. (eds.) Intelligent Music Information Systems – Tools and Methodologies, pp. 147–176. Information Science Reference – formerly Idea Group (2008)
9. Mierswa, I., Wurst, M.: Efficient case based feature construction for heterogeneous learning tasks. In: Gama, J., Camacho, R., Brazdil, P.B., Jorge, A.M., Torgo, L. (eds.) ECML 2005. LNCS (LNAI), vol. 3720, pp. 641–648. Springer, Heidelberg (2005)
10. Mierswa, I., Wurst, M., Klinkenberg, R., Scholz, M., Euler, T.: YALE: Rapid Prototyping for Complex Data Mining Tasks. In: Proceedings of the 12th ACM SIGKDD International Conference on Knowledge Discovery and Data Mining (KDD 2006). ACM Press, New York (2006)
11. Mörchen, F., Ultsch, A., Thies, M., Löhken, I., Noecker, C., Stamm, M., Efthymiou, N., Kuemmerer, M.: Musicminer: Visualizing perceptual distances of music as topograpical maps. Technical report, Department of Mathematics and Computer Science, University of Marburg, Germany (2004)
12. Pohle, T., Pampalk, E., Widmer, G.: Evaluation of frequently used audio features for classification of music into perceptual categories. In: Proceedings of the International Workshop on Content-Based Multimedia Indexing (2005)
13. Strehl, A., Ghosh, J.: Cluster ensembles – a knowledge reuse framework for combining partitionings. In: Proceedings of AAAI 2002, Edmonton, Canada (2002)
14. Tzanetakis, G.: Manipulation, Analysis and Retrieval Systems for Audio Signals. PhD thesis, Computer Science Department, Princeton University (2002)

15. Wurst, M.: Distributed Collaborative Structuring – A Data Mining Approach to Information Management in Loosely-Coupled Domains. PhD thesis, LS8, Faculty of Computer Science, TU Dortmund, Germany (2008)
16. Wurst, M., Morik, K.: Distributed feature extraction in a p2p setting - a case study. Future Generation Computer Systems, Special Issue on Data Mining (2006)
17. Wurst, M., Morik, K., Mierswa, I.: Localized alternative cluster ensembles for collaborative structuring. In: Proceedings of the European Conference on Machine Learning (2006)
18. Zils, A., Pachet, F.: Automatic extraction of music descriptors from acoustic signals using eds. In: Proceedings of the 116th Convention of the AES (2004)

Micro Information Systems
and Ubiquitous Knowledge Discovery

Rasmus Ulslev Pedersen

Copenhagen Business School,
Dept. of Informatics, Embedded Software Lab,
Copenhagen, Denmark
rup.inf@cbs.dk

Abstract. The combination of two new fields is introduced: a field named *micro information systems* (micro-IS) and the field named ubiquitous knowledge discovery (KDubiq). Each of the fields offer new ways to understand the world of electronic devices that surround us. The micro-IS field primarily focuses on the combination of embedded systems and information systems, while the KDubiq system focuses on extracting the knowledge that is available from these systems. Furthermore, we demonstrate the LEGO®MINDSTORMS®NXT platform, and discuss the potential usefulness of running TinyOS, Squawk, and .NET Micro Framework using this platform.

1 Introduction

The insight we gain from this study is partly based on the KDubiq framework, as described in the first part of this book, and the micro-IS field together. This paper discusses challenges that emerge from combining the field of *micro information systems* with the field of ubiquitious knowledge discovery (KDubiq). The KDubiq vision generates some interesting challenges [3] in the micro-IS field. For example, it is desirable to extract and combine raw data to form information from micro-IS systems, but due to certain system properties such as a small processor and battery limitations, this can be a difficult task.

The research field of *information systems* (IS) has changed only slowly: IS researchers have mainly focused on server, web, and PC-oriented applications. The reason for this focus has probably been that the power of the server and PC has followed Moore's law and therefore created many new possibilities. But in parallel and mostly unnoticed in IS, the market for embedded systems has surged. Today, more than 90% of all processors are not server- and PC- central processing units (CPU), but micro control processors (MCU). These MCUs are embedded in cars, sensors, doors, watches, mobile phones, alarm systems, etc. So there is already a large set of applications and the IS community does not need to look for new applications to understand the implications. However, there is a massive bulk of research that awaits in the new micro information system

M. May and L. Saitta (Eds.): Ubiquitous Knowledge Discovery, LNAI 6202, pp. 216–234, 2010.

sub-field. Most of the existing work in IS can be reflected and re-analyzed in terms of micro information systems.

In the recent decades most machine learning and data mining research has been rooted in larger server systems. The results have mainly been targeted at providing information for humans sitting in front of a monitor. However, in recent years there has been more focus on the smaller systems. Given that the vast majority of processors are embedded in some system, this move is not surprising. The KDubiq vision adds a direction and provides some interesting and open challenges to move further in this direction.

We show a preliminary idea of what micro information systems are by showing a classification of micro information systems. We introduce both mobile- and nano- information systems to show the two end-points in the micro information systems field. Furthermore, we show the stationary IS systems, by which we mean server- and PC-based information systems. The reason for choosing size and degree of customization as the two ways to classify micro information systems are because these micro information systems will be designed for a particular purpose on a case-by-case basis. This is not to say that stationary information systems are production line work, but simply acknowledging that these kind of larger information systems are some decades older than micro information systems thus requiring less customization from one application to the next. See Figure 1 for this introductory classification.

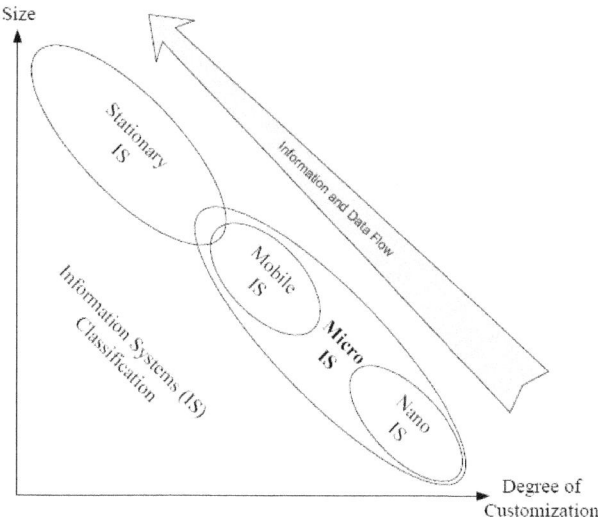

Fig. 1. Micro information systems classification and information flow

In this context, we present the micro-IS components and then we add the KDubiq aspect to each component to see potential synergies. The organization is such that we start with a theoretical definition of a micro information system in Section 2. Then we go on to show some small prototypes of mobile-, micro-, and nano-information systems. After the theory section, we show how one can use a certain platform for micro-IS experimentation. This takes place in Section 4. The hardware platforms suggested are mainly LEGO MINDSTORMS NXT and the Sun Spot from Sun Micro Systems. The operating systems examples are mainly the embedded Squawk Java Virtual Machine, TinyOS, and the .NET Micro Framework from Microsoft. In Section 5 we discuss the new LEGO MIND-STORMS NXT platform in more detail, which is on its way to become a popular learning platform for educational and research purposes. With the platform description in mind, we go on to discuss the requirements for a micro-IS platform as seen from the viewpoint of IS students, teachers, industry, and researchers. This takes place in Section 6. Some of the teaching experience with the system is presented in Section 7. In order so show that this idea about micro information systems is real, we include a summary section on selected experiences from a European research summer school and at a business school for freshmen students; see Section 6, which also covers a requirements analysis from different user groups of micro-IS. At the end, we conclude that micro-IS with KDubiq is promising and points to future work.

2   Micro Information Systems

In this section we will combine micro-IS with KDubiq. We provide the following micro-IS definition:

Micro information systems are embedded information systems subject to additional significant contextual constraints related to size, form, and function.

It is a new area, and work remains to further define how this new field can support the IS research community such as the mission of the Association for Information Systems (AIS)[1]. In addition, we have outlined the major relevant fields together with KDubiq in Figure 2.

The different research areas in Figure 2 are included to show some main areas that micro-IS will draw on. Information systems is the main area providing most of the new research questions. We can imagine contexts like business informatics, wireless sensor networks, ubiquitous/pervasive computing, etc. The type of information in a micro-IS is different than a traditional IS, so we include information theory and machine learning as two important areas. Finally, embedded systems provides the wireless sensor network experimental platform discussed in Section 4.1.

To justify the novelty of micro-IS, we can consider these arguments: It is new because it brings the concept of something *small* (i.e. embedded) to the IS field,

---

[1] The mission of the Association for Information Systems is "to advance knowledge in the use of information technology to improve organizational performance and individual quality of work life."

it explicitly adds a new kind of information, and it certainly focuses on a new kind of platform, which is much smaller than what we are used to consider. It can be argued that it is similar to the pervasive/ubiquitous visions, but we think they are mainly visions, with somewhat limited practical resemblance to IS. By defining the new micro-IS field to be within IS, the research community can decide exactly how the artifacts (i.e. micro systems) are analyzed and observed, which we think is the optimal approach. The *information* we examine is also different. It comes in smaller quantities and can be as small as a binary value with some inherent significance. To look at information in its most basic form is refreshing, and it is also why we have put information theory up as a relevant toolbox to take into account.

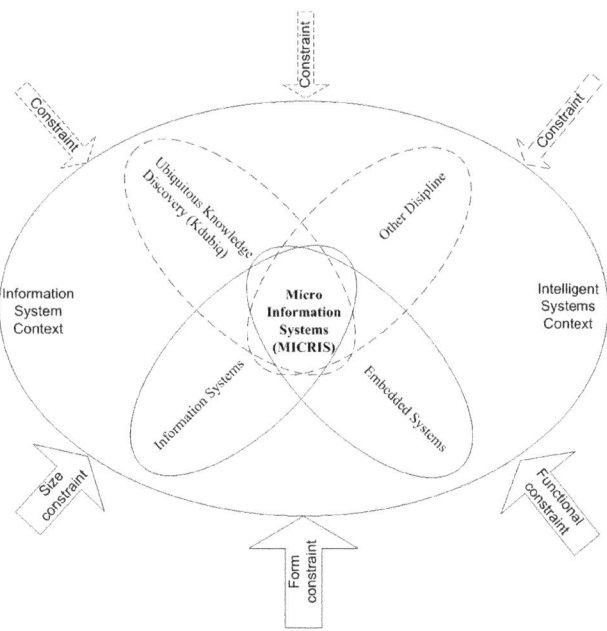

Fig. 2. micro-IS systems are surrounded by other research fields and extended to include the KDubiq aspect

The KDubiq addition to Figure 2 show this field interacts with the core micro-IS disciplines of embedded systems and information systems. It is in this context that we must extract knowledge from the system and initiate the flow and transformation of data to information toward the larger IS systems in the context. This flow is also illustrated in Figure 1.

2.1   Micro Information Systems Examples

Defining an area called micro information systems (micro-IS) adds new perspectives to the IS field. Micro-IS systems are often based on information such as temperature and registrations of movements while traditional information in IS systems could be database records containing some product ID or a quantity. Furthermore, it introduces a new kind of *system*; namely one that is physically designed toward its application. The novelty lies in the fact that it is not only a software-oriented, but equally a hardware-oriented design. Finally, a micro-IS system can be analyzed from many different perspectives:

Research unit: Individual, group, or organizational level.
Development: Software and hardware development perspectives.
Information management: Databases and information retrieval.
Standards: Open source and security.
Applications: ERP, CRM, and business intelligence systems.

There is a small demo of a micro-IS system in Figure 3: Refer to the (work-in-progress) Sourceforge project *nxtmote* and the *nxtsquawk* project at dev.java.net. It is developed using the Eclipse *nescdt* editor plugin for TinyOS.

The closest neighbor to micro-IS when looking at larger systems is mobile phones as depicted in Figure 1: It shows that micro information systems are smaller than stationary information systems and require more customization on a case-by-case basis. Furthermore, it shows how data and information mostly flow toward the larger systems. Mobile phones are also small specialized units that use (powerful) micro processors, and they are the subject of the next section.

A micro information system is also an embedded system. To extract the data and potential knowledge from it, we can draw on the fact (as mentioned above) that hardware design and software design go hand in hand for these kind of systems. So if the KDubiq value proposition is strong enough, then it can make sense to upgrade the hardware on a given platform, such that it becomes possible to run advanced algorithms on it. We also mention business processes like outsourcing above. This is a more challenging thing to do in a KDubiq context. While there is no significant problem in outsourcing software or even hardware development, there is probably not yet any tradition for contracting away the development of the intelligent algorithms. An example research question when looking at KDubiq from a micro-IS perspective, could be: How do we provide the right amount of knowledge given the micro-IS constraints of size, form, and function? That is clearly a challenge to balance these constraints.

2.2   Mobile Information Systems

There is nothing new in seeing a mobile phone as an information system. However, there are still many traditional IS aspects that remain largely untouched on mobile phones, such as outsourcing hardware and software development.

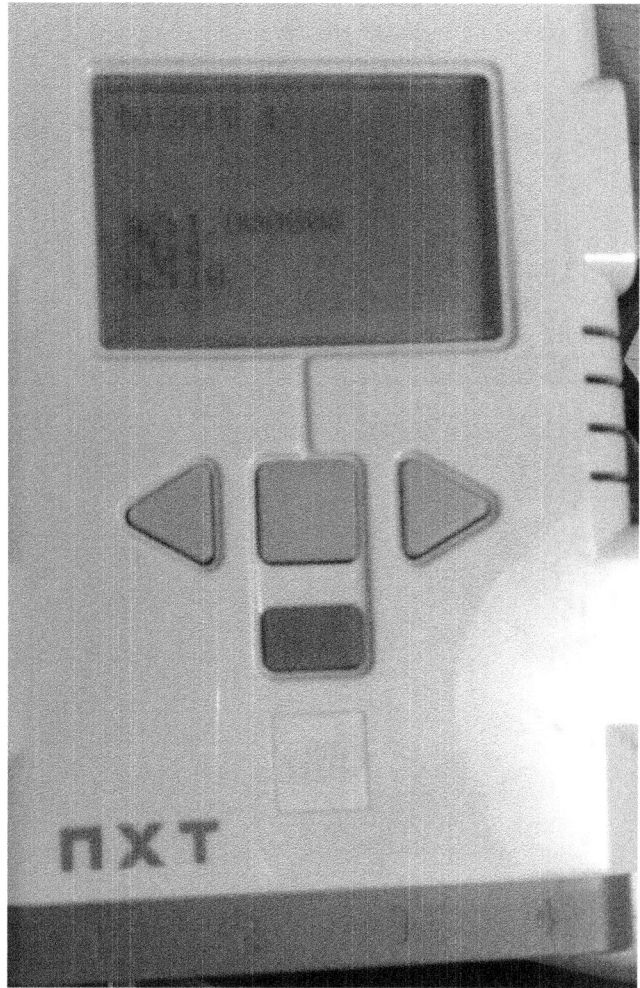

Fig. 3. Micro Information System (micro-IS). Developed using LEGO MINDSTORMS NXT.

The reason that we introduce mobile information systems (MOBIS) is that it will be an important interface device within micro information systems. See Figure 4 for a small demo. It is developed using the Google Android software development kit (SDK) and the Android Eclipse Plugin. Notice that the Google Android runs an application that displays *Mobile Information System 1* on the screen.

The KDubiq framework fits very well into mobile information systems. There are several factors that contribute to this fact. The mobile platforms are ubiquitous. They provide a consistent software interface such as the Java Micro Edition or the platform specific software development kits (SDK) like the Google Android

SDK. In addition, the power model is flexible for mobile information systems as they provide easy access to recharge the battery.

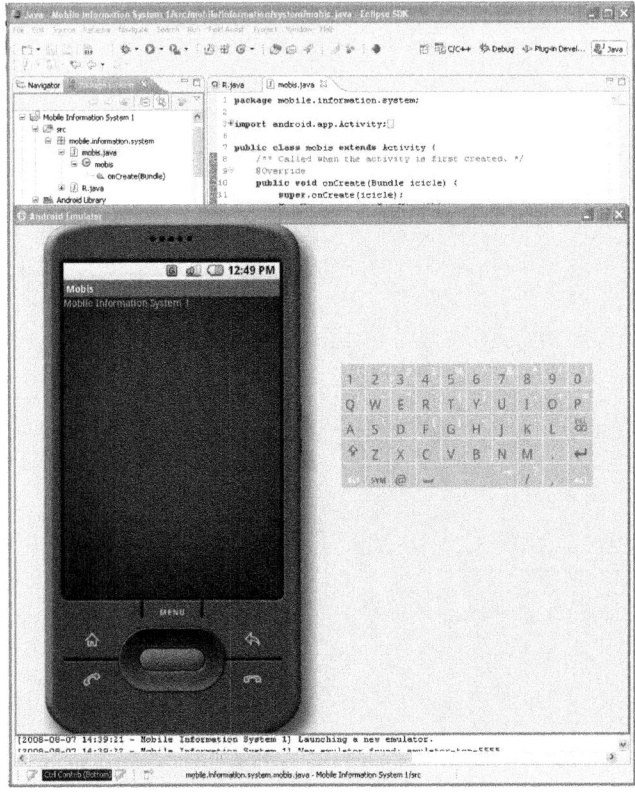

Fig. 4. Mobile Information System

2.3   Nano Information Systems

We introduce this kind of information system for two reasons: first and foremost it serves the purpose of indirectly defining micro information systems, as we define nano information systems (nano-IS) as an overlapping field to micro-IS. The second reason for defining nano-IS is simply because we believe that it will be highly important in the IS field, starting in a few years.

A small type of micro-IS is nano information systems: Figure 5 shows a program that allow some small scale technology experimentation. To show that the idea of nano-IS is perhaps not far away, we created a small project with a few atoms (the details are not important), and called the system *Nano Information System 1*, as we believe that it is the first idea to define a nano information system in the IS field. From Figure 5 it can be seen the weight of the system is around

10^{-24} kg, so it is very light!. Developed using the Eclipse Rich Client (RCP) framework-based application nanoXplorer IDE from nanoTitan. The system is labeled *Nano Information System 1*. It is off course just an artificial example, but the point is that nano information systems are very much smaller than what we normally focus on.

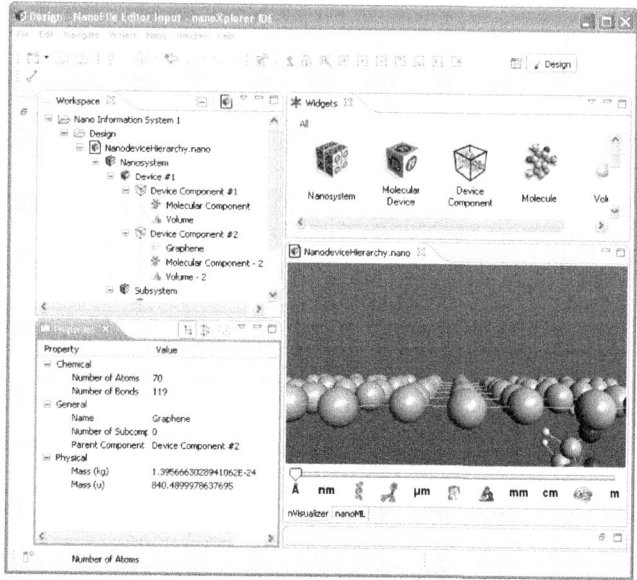

Fig. 5. Nano Information System

To name a few research questions in the nano-IS end of micro-IS, we could think about how HCI works when you can't see or feel a sensor, as could be the case of nanosensors[2] As for systems development, we can ponder questions like how we develop, update, and maintain a nano system.

It is a new way to see KDubiq in relation to nano information systems. However, in the future we will get access to all sorts of interesting data from nano information systems. This will provide (yet) another set of challenges for KDubiq, and pave the way for even more interesting and useful research projects. For example, with a nano information system embedded into humans, we could perhaps better be able to determine when a certain virus is about to spread as an epidemic.

Now that we have covered mobile-, micro-, and nano- information systems to some extent, we will introduce the experimental platform that can be used for micro-IS research.

---

[2] See http://en.wikipedia.org/wiki/Nanosensor for a definition of nanosensor.

3   KDubiq with Micro Information Systems

The KDubiq model includes elements from different areas of data mining and ubiquitous research. Specifically, the focus has been on

- Application,
- Ubiquitous,
- Resource-aware,
- Data Collection,
- Privacy and Security, and
- HCI and User.

These defining KDubiq elements provide the major projections of the framework. In the previous section, we provided several examples of how micro information systems could materialize. In the present section, we will map the KDubiq elements to the proposed micro information system framework.

Application: A KDubiq application can range from distributed intelligent music preference recommendation to intelligent cars. A micro-IS application is also part of a larger information system context and data can flow freely in the system.

Ubiquitous: One characteristic of the KDubiq system is the distributed (i.e. ubiquitous) property. This is shared with a micro-IS, which also are found everywhere.

Resource-aware: The resource-awareness is shared between a micro-IS and a KDubiq system for many devices. If a device is battery-powered it is clear the intelligent algorithms will need to be adapted to obey a demand for long operating times.

Data Collection: The KDubiq ideas is the data is gradually refined as it moves from one distributed node to the next. In micro-IS it is similar, and and the smaller system provide sensor input which is treated with machine learning algorithms and possibly later used for actuator feedback.

Privacy and Security: It is a challenge in a KDubiq system to keep the data private. It is not only a traditional security challenge, but a challenge of finding ways to data mine without revealing sensitive data. In a micro-IS this would fall under a broader ethical consideration.

HCI and User: A micro-IS is an embedded information system. It usually lacks the traditional HCI elements like a big screen and a full-size keyboard. Instead the embedded system has to act intelligently and autonomously to compensate for the restricted user interface. Many KDubiq projects would encompass such embedded elements.

The KDubiq is defined for problems which have strong distributed and intelligence components. The micro-IS examples provided in Section 4.1 provides an example of wireless sensor networks. It is a good example of a KDubiq platform where the elements of applications, ubiquitousness, resource-awareness, data collection, privacy/security, and HCI to a strong extent are very relevant. Each of the aforementioned elements are important for wireless sensor networks.

In the following section, we provide one such example of a prototyping platform for KDubiq experiments. It is a small embedded system with can be used to collect and process distributed sensor data.

4   Experimental Platform

It is valuable to be able to experiment with micro-IS systems, and therefore we suggest to define the area such that most experiments and research work consists of both a theoretical part and an experimental part.

Here we can look to a successful research field called wireless sensor networks (see `tinyos.net`). This community has a tradition for accompanying research work by a demo system.

4.1   Micro Information Systems and Wireless Sensor Network

The recent introduction of LEGO MINDSTORMS NXT[3] paved the way for micro-IS (see Figure 6). We will discuss how this platform can be helpful across a number of requirements in terms of teaching.

We propose LEGO MINDSTORMS as a standard platform and we show one example of how it can be used together with wireless sensor networks. The system consists of TinyOS, Squawk and LEGO MINDSTORMS. In earlier work [5] it was necessary to use several different platforms to try embedded Java and TinyOS. With LEGO MINDSTORMS we can use the same hardware platform all the time.

TinyOS [2] is a small operating system for wireless sensor networks. We have described a port of TinyOS for educational purposes [6]. It runs easily on the NXT hardware. Originally the operating system ran on motes like the Mica2, as shown in Figure 7: Sun SPOT is the recent WSN platform from Sun Microsystems. The CCZACC06 (now CC2480) is a recent Zigbee radio from Texas Instruments. The CC2420 is widely in use today within WSN. The operating system is small and there is plenty of room for drivers and other software on the main processor in NXT. We also have a website, `http://nxtmote.sf.net`, for the first port of TinyOS to NXT.

Squawk [1] is a research Java Virtual Machine (JVM) from Sun, which was released in the spring of 2008. It is the JVM that runs inside the recently released wireless sensor network platform *SUN SPOT* (see Figure 7). An ongoing project for setting up the Squawk Java virtual machine from Sun Microsystems on NXT can be found at `https://nxtsquawk.dev.java.net/`.

There are some interesting projects on their way: With the Microsoft .NET Micro Framework hardware portability kit is possible to fit the image onto devices with 256 KB flash and 64 KB ram. This coincides with the current LEGO

---

[3] See `http://mindstorms.lego.com` for a full description of NXT.

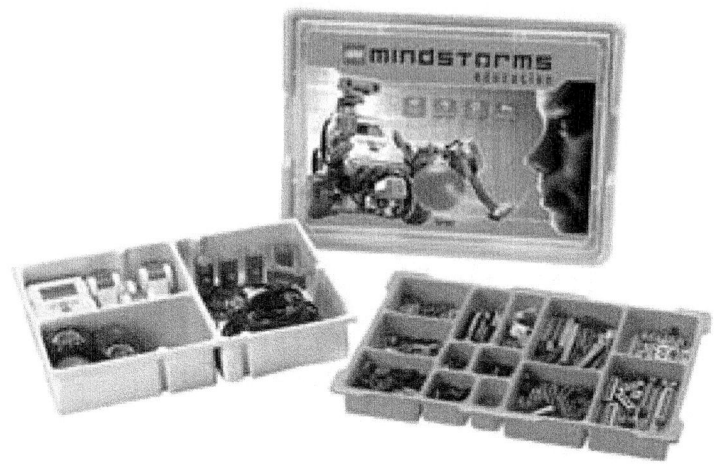

Fig. 6. The LEGO MINDSTORMS NXT educational kit

(a) Mica2 Mote   (b) Sun Spot

(c)   Zigbee   Radio   (d) 802.15.4   Transceiver
(CCZACC06)   (CC2420)

Fig. 7. Mica was one of the original platforms for running TinyOS motes

NXT platform and a project called *nxtdot* (see Figure 8) has been initialized on Sourceforge. Pressing the orange button outputs the micris demo text to the display. The code can be accessed at http://nxtdot.sf.net.

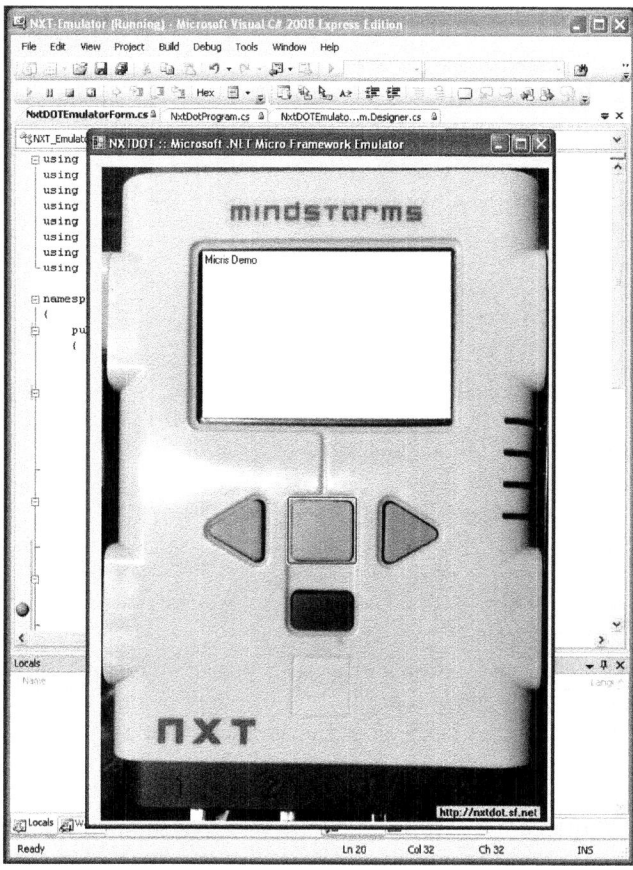

Fig. 8. An Emulated NXT with Microsoft .NET Micro Framework 3.0

Some of the data-related KDubiq challenges (see Chapter 4) demand a different set of skills than only programming for the PC and servers. We believe that the wireless sensor networks provide a good testbed for KDubiq projects. The processors on these platforms are often small and they are always battery driven.

4.2   Support Vector Machine Experiment

The experimental platform, LEGO MINDSTORMS NXT, has been used at the KDubiq summer school. It was used to classify a data point on one platform

Listing 1.1. TinyOS nesC code

```
module HalBtM {
  provides interface HalBt;
  uses interface HplBc4;
}
implementation {
  event void Boot.booted() {
    call BTIOMapComm.setIOMapCommAddr(&IOMapComm);
      call BTInit.init ();
      cCommInit(NULL);
  }
  command error_t HalBt.sendMsg(uint8_t* msg){
      error_t  error  = SUCCESS;
    return error ;
  }
  ...
```

which was collected on a different platform. The first NXT would collect a two-dimensional data point from a light sensor and a ultra sound distance sensor. Then this datapoint would be transferred to a second NXT, which then classified the point according to the trained SVM.

The classification of the binary SVM:

$$f(\boldsymbol{x}, \boldsymbol{\alpha}, b) = \{\pm 1\} = sgn\left(\sum_{i=1}^{l} \alpha_i y_i k(\boldsymbol{x}_i, \boldsymbol{x}) + b\right) \tag{1}$$

A collected vector, $\boldsymbol{x}$ is classified in to a positive class $+1$ or negative class -1 based on the values of the Lagrange multipliers α for the support vectors.

In the experiment, one NXT uses the `HalBt.sendMsg` (Listing 1.1) to send the data point to be classified to the other NXT.

The SVM in the exercise makes a dot product with two vectors as shown in Listing 1.2. The vectors $\boldsymbol{a}$ and $\boldsymbol{b}$ are used in the *dotk* method to reach the result.

The platform, LEGO MINDSTORMS NXT, is introduced in the next section.

5   LEGO MINDSTORMS NXT

Established and emerging embedded operating systems can benefit from the implementation of LEGO MINDSTORMS NXT for several reasons. The open source policy of the recently released NXT makes it easy to import new operating systems to NXT. NXT can be seen in Figure 9 and 10. The sensors can be seen in Figure 11.

The MINDSTORMS NXT system from LEGO Corp. is an interesting and flexible hardware platform. The NXT PCB (the green plate in Figure 10) is equipped with an ARM MCU as well as an AVR MCU; both MCUs are from Atmel. Furthermore, these two popular MCUs are connected to input ports, output ports, an LCD, Bluetooth radio, and USB. Moreover, there is already

Listing 1.2. SVM TinyOS nesC code

```
module TinySVMM {

  provides interface TinySVM as TSVM;

  uses interface FixedPoint as FP;

}
implementation {
    // test data
    dmfp tsd;
    // test data label
    dvfp tsy;
...
    // make dot product
    command fp_t TSVM.dotk(fp_t a, fp_t b){
      fp_t r;
      r = call FP.mulfp(a,b);
      return r;
    }
}
```

a rich set of sensors available from both LEGO and third-party vendors, which enables the use of the NXT system for prototyping in relation to almost any conceivable education (or research) project.

Fig. 9. A NXT with 4 input sensors (touch, sound, light and ultrasonic) and the 3 motors are also shown

From an educational perspective, the number of different sensors is excellent. Many of the sensors shown in this section are very affordable and easy to obtain. This means that machine learning for educational purposes can begin to include the physical world in its experiments in a way that has never before been possible on a international scale (LEGO is available almost everywhere).

There are many platforms available that could be used in KDubiq projects. One can choose from a vast selection of hardware platforms running everything from Linux to Java to .NET. However, one reason to use the LEGO MIND-STORMS NXT platform is that it is becoming widespread both in industry and in academia.

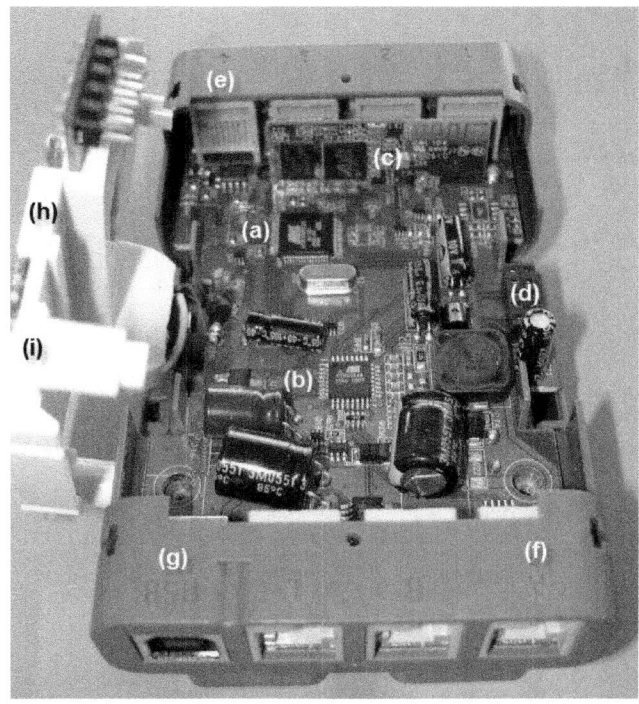

Fig. 10. NXT with important hardware highlighted: (a) ARM7 MCU, (b) ATmega48 MCU, (c) CSR BlueCore4 Bluetooth radio, (d) SPI bus and touchpad signals, (e) high-speed UART behind input port 4, (f) output (generally motor) port, (g) USB port, (h) four-button touchpad, and (i) 100x64 LCD display

5.1   Sensors

A micro-IS platform is connected to the real physical world, so the short discussion of the sensors is well placed here. The input and output ports feature a 6-wire RJ12 connector. On the input ports there are both analog and digital lines. On the output ports, pulse width modulation is used with the motors in the standard NXT firmware. The NXT comes with a basic set of sensors. This basic set includes an ultrasonic distance measurement sensor, a light intensity sensor, a sound sensor, a touch sensor, and motors. Moreover, there are a number of third-party sensors available such as various acceleration sensors, a compass sensor, a temperature sensor, etc.

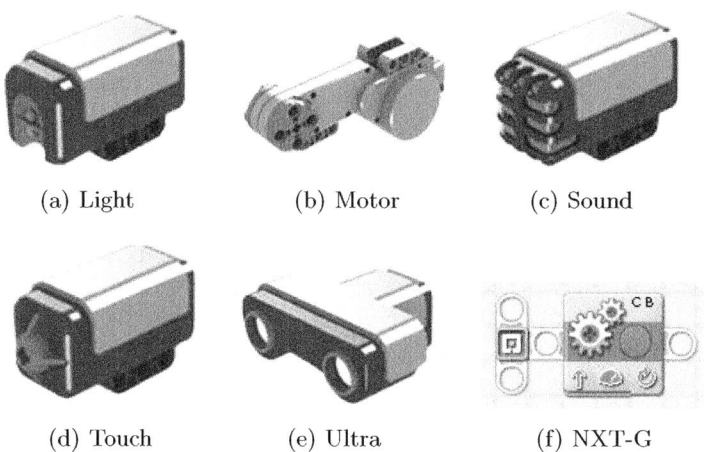

(a) Light   (b) Motor   (c) Sound

(d) Touch   (e) Ultra   (f) NXT-G

Fig. 11. Standard LEGO MINDSTORMS NXT sensors and NXT-G block

With NXT comes a set of standard sensors, as shown in Figure 11. This set of sensors can give most students enough to work with in order to create their custom motes: the light and microphone sensor are almost standard for a mote. It should not go unnoticed that there are three motors (see Figure 11b), which makes it simple to create dynamic moving and data collecting motes. For the sake of completion, there is one NXT-G block shown, which is used in the block-based programming language.

For KDubiq, the sensor inputs are especially interesting. It is from the sensors that the primary raw data comes from, and it is a main source of ubiquitous data from which the knowledge discovery takes place. We should note that embedded systems also features actuators, which are devices that "do something" to the world. However, from a KDubiq and micro-IS perspective, we think the sensors are the main sources of information. When this information is collected across a KDubiq problem, then the information can be used to trigger actuators as a result.

One of the advantages of multiple communities sharing the NXT hardware and sensors is that mass-production drives prices down to a level where classroom teaching becomes quite affordable. The technical discussion above now has to be be mapped to the different users of an micro-IS platform. This is the subject of the next section.

6 Requirements Analysis from Research and Educational Micro-IS Users

In this section we look at the requirements that different groups of people have in terms of using LEGO MINDSTORMS NXT for researching and teaching micro-IS and the embedded part of KDubiq projects. We argue that the LEGO MINDSTORMS NXT is presently the most relevant hardware platform for this purpose.

6.1 Students

A broad definition of a *student* is used. It stretches all the way from the first year at university or college to finishing a Ph.D. and becoming a teacher. We believe that students would like to work with LEGO MINDSTORMS NXT because LEGO is well-known to many students around the world. They immediately recognize the appealing looks of the NXT system and the familiarity of the bricks and connectors. This means that no time has to be invested in learning a proprietary system before the first inventions can be created. Furthermore, the students recognize the value of knowing LEGO MINDSTORMS will stretch beyond the classroom setting and on to family life and hobbies. A different platform featuring a real-time Java processor has been used at Copenhagen Business School by the authors of these papers [4][7].

6.2 Teachers

From a teacher standpoint the system should be useful across many subjects. In addition, the systems used would ideally be portable to other institutions to make time investments worthwhile. Due to the ubiquitous use of LEGO MIND-STORMS (at least compared to other educational systems) the teacher would immediately become part of a well-functioning community.

6.3 Industry

It is best for industry if the student learns using a system that prepares them for real applications. It is equally important for industry to have a system that is able to upgrade the skillset of existing employees. The field of embedded machine learning is just developing and anyone not in school or able to attend conferences will miss the chance to catch on to this rewarding area. LEGO MINDSTORMS NXT is available on many websites and the employee or the employer can choose to invest in a few sets to get started.

6.4 Researchers

The research role is often associated with a corresponding teaching obligation. If that is the case then the researcher would be looking for ways to combine teaching efforts with research efforts. With LEGO MINDSTORMS NXT this is possible. Many schools are able to acquire a number of NXT sets, and due to the open hardware/software source nature of LEGO MINDSTORMS the researcher can use the same hardware in the laboratory as in the classroom.

7   Teaching Experiences with TinyOS and LEGO MINDSTORMS NXT

We have tried the TinyOS and LEGO MINDSTORMS NXT combination with different groups. The youngest group was a second-semester business/computer science profile. It was possible in 5 hours total during a theme week with LEGO MINDSTORMS to get radio connection between two units and manipulate the display. The students were very excited, even though we initially thought that TinyOS would be much too hard for such a young group. No actual programming was conducted in that short period.

The second group we have been able to expose the system to was 30-40 masters/Ph.D. students at a large EU project on ubiquitous knowledge discovery (KDubiq) (see Figure 12). The teaching part of introducing TinyOS, the associated programming language nesC, the software, and hardware parts of LEGO

(a) 18 NXT in one bag   (b) KDubiq Summer School

(c) Teaching   (d) Exercises

Fig. 12. Teaching TinyOS on LEGO MINDSTORMS at the KDubiq Summer School for about 2 hours plus 4 hours of labs. See http://www.kdubiq.org for more.

MINDSTORMS NXT took 2 hours. Then 4 hours of laboratory work followed where each group had two sets of LEGO MINDSTORMS NXTs. The task was to use an algorithm on one NXT and send the results to another NXT via Bluetooth.

8   Conclusion

Micro Information Systems in combination with ubiquitous knowledge discovery will probably play a significant role in the IS community for years to come. KDubiq can potentially provide a framework for understanding the complicated processes of extracting knowledge from distributed heterogenous systems both in terms of hardware/software and also in terms of the complex data structures in such systems.

In this study we have observed that LEGO MINDSTORMS NXT is one possible platform for micro-IS learning and research. Several examples of early use of this system has been provided ranging from first year students to master students and Ph.D. students. We argue the this platform is excellent for experimental KDubiq work and related challenges [3].

References

1. Horan, B., Bush, B., Nolan, J., Cleal, D.: A platform for wireless networked transducers. Technical Report TR-2007-172, Sun Microsystems (2007)
2. Levis, P., Madden, S., Gay, D., Polastre, J., Szewczyk, R., Whitehouse, K., Woo, A., Gay, D., Hill, J., Welsh, M., Brewer, E., Culler, D.: Tinyos: An operating system for sensor networks. In: Weber, W., Rabaey, J., Aarts, E. (eds.) Ambient Intelligence. Springer, Heidelberg (2004)
3. May, M., Berendt, B., Cornéjols, A., Gama, J., Gianotti, F., Hotho, A., Malerba, D., Menesalvas, E., Morik, K., Pedersen, R., Saitta, L., Saygin, Y., Schuster, A., Vanhoof, K.: Research challenges in ubiquitous knowledge discovery. In: Kargupta, H., Han, J., Yu, P.S., Motwani, R., Kumar, V. (eds.) Next Generation of Data Mining, ch. 7, pp. 131–150. Chapman & Hall/CRC (2008)
4. Pedersen, R., Schoeberl, M.: An embedded support vector machine. In: Proceedings of the Fourth Workshop on Intelligent Solutions in Embedded Systems (WISES 2006), Vienna, Austria, pp. 79–89 (June 2006)
5. Pedersen, R.U.: Using Support Vector Machines for Distributed Machine Learning. PhD thesis, Dept. of Computer Science, University of Copenhagen (2005)
6. Pedersen, R.U.: Tinyos education with lego mindstorms nxt. In: Gama, J., Gaber, M.M. (eds.) Learning from Data Streams. Processing Techniques in Sensor Networks, ch. 14, pp. 231–241. Springer, Heidelberg (September 2007)
7. Schoeberl, M., Pedersen, R.: WCET analysis for a Java processor. In: Proceedings of the 4th International Workshop on Java Technologies for Real-time and Embedded Systems (JTRES 2006), pp. 202–211. ACM Press, New York (2006)

MineFleet®: The Vehicle Data Stream Mining System for Ubiquitous Environments

Hillol Kargupta, Michael Gilligan, Vasundhara Puttagunta, Kakali Sarkar, Martin Klein, Nick Lenzi, and Derek Johnson

Agnik, LLC, 8840 Stanford Blvd. Suite 1300, Columbia, MD 21045, USA
{hillol,mgilligan,kakali}@agnik.com
http://www.agnik.com

Abstract. This paper describes the MineFleet® distributed vehicle performance data stream mining system designed for commercial fleets. The MineFleet Onboard analyzes high throughput data streams onboard the vehicle, generates the analytics, and sends them to the remote server over the wireless networks. The paper describes the overall architecture of the system, business needs, and shares experience from successful large-scale commercial deployments. MineFleet is probably one of the first distributed data stream mining systems that is widely deployed at the commercial level. The paper discusses an important problem in the context of the MineFleet® application—computing and detecting changes in correlation matrices in a resource-contrained device that are typically used onboard the vehicle. The problem has immediate connection with many vehicle performance data stream analysis techniques such as principal component analysis, feature selection, and building predictive models for vehicle subsystems.

1 Introduction

The wireless and mobile computing/communication industry is producing a growing variety of devices that process different types of data using limited computing and storage resources with varying levels of connectivity to communication networks. The rich source of data from the ubiquitous components of businesses, mechanical devices, and our daily lives offers the exciting possibility of a new generation of data intensive applications for distributed and mobile environments. Mining distributed data streams in an ubiquitous environment is one such possibility. Several years of research on distributed data mining [5,7] and data stream mining have produced a reasonably powerful collection of algorithms and system-architectures that can be used for developing several interesting classes of distributed applications for lightweight wireless applications. In fact an increasing number of such systems [4,8] are being reported in the literature. Some commercial systems are also starting to appear.

This paper reports the development of MineFleet®, a novel mobile and distributed data mining application for monitoring vehicle data streams in real-time. MineFleet is designed for monitoring commercial vehicle fleets using

M. May and L. Saitta (Eds.): Ubiquitous Knowledge Discovery, LNAI 6202, pp. 235–254, 2010.

onboard embedded data stream mining systems and other remote modules connected through wireless networks in a distributed environment. Consider a nationwide grocery delivery system which operates a large fleet of trucks. Regular maintenance of the vehicles in such fleets is an important part of the supply chain management and normally commercial fleet management companies get the responsibility of maintaining the fleet. Fleet maintenance companies usually spend a good deal of time and labor in collecting vehicle performance data, studying the data offline, and estimating the condition of the vehicle primarily through manual efforts. Fleet management companies are also usually interested in studying the driving characteristics for a variety of reasons (e.g. policy enforcement, insurance). Monitoring fuel conumption, vehicle emissions, and identifying how vehicle parameters can be optimized to get better fuel economy are some additional reasons that support ample return of investment (ROI) for systems like MineFleet.

The MineFleet is widely adopted in the fleet management industry. Similar applications also arise in monitoring the health of airplanes and space vehicles. There is a strong need for real-time on-board monitoring and mining of data (e.g. flight systems performance data, weather data, radar data about other planes). The MineFleet system can also be applied to this aviation safety domain where it monitors planes and space vehicles instead of automobiles and trucks.

The main unique characteristics of the MineFleet system that distinguish it from traditional data mining systems are as follows:

1. Distributed mining of the multiple mobile data sources with little centralization of the data.
2. Onboard data stream management and mining using embedded computing devices.
3. Designed to pay careful attention to the following important resource constraints:
   (a) Minimize data communication over the wireless network.
   (b) Minimize onboard data storage and the footprint of the data stream mining software.
   (c) Process high throughput data streams using resource-constrained embedded computing environments.
4. Respect privacy constraints of the data, whenever necessary.

Section 2 presents an overview of the MineFleet system. Section 3 describes the need for monitoring the correlation matrices in MineFleet. Section 4 describes the computing problem. Sections 5 and 6 describe the algorithm. Section 7 reports some experimental results. Finally, Section 8 concludes this paper.

2   MineFleet®: An Overview

MineFleet®is a mobile and distributed data stream mining environment where the resource-constrained " small" computing devices need to perform various non-trivial data management and mining tasks on-board a vehicle in real-time.

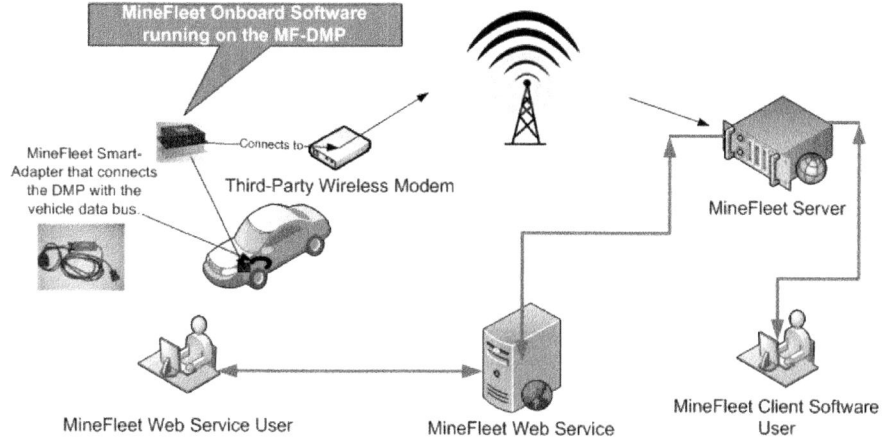

Fig. 1. MineFleet architecture. © Copyright, Agnik, LLC.

MineFleet analyzes the data produced by the various sensors present in most modern vehicles. It continuously monitors data streams generated by a moving vehicle using an on-board computing device, identifies the emerging patterns, and if necessary reports these patterns to a remote control center over low-bandwidth wireless network connection.

MineFleet also offers different distributed data mining capabilities for detecting fleet-level patterns across the different vehicles in the fleet. This section presents a brief overview of the architecture of the system and the functionalities of its different modules.

The current implementation of MineFleet analyzes and monitors only the data generated by the vehicle's on-board diagnostic system and the Global Positioning System (GPS). MineFleet Onboard is designed for embedded in-vehicle computing devices, tablet PCs, and cell-phones. The overall conceptual process diagram of the system is shown in Figure 1. The MineFleet system is comprised of four important components:

1. On-board hardware: MineFleet Onboard module is comprised of the computing device that hosts the software to analyze the vehicle-performance data and the interface that connects the computing device with the vehicle data bus (OBD-II for light duty and J1708/J1939 for heavy duty vehicles). Figure 2 shows the MineFleet Onboard Data Mining platform (MF-DMP101) device that hosts the MineFleet Onboard software. MineFleet also runs on many different types of embedded devices, in-vehicle-tablet-PCs, laptops, cellphones and other types of handheld devices.

2. On-board data stream management and mining module: This module manages the incoming data streams from the vehicle, analyzes the data using various statistical and data stream mining algorithms, and manages the transmission of the resulting analytics to the remote server. This module

Fig. 2. MineFleet Data Mining Platform (MF-DMP101) that hosts the MineFleet On-board software. © Copyright, Agnik, LLC.

also triggers actions whenever unusual activities are observed. It connects to the MineFleet Server located in a data center through a wireless network. The system allows the fleet managers to monitor and analyze vehicle performance, driver behavior, emissions quality, and fuel consumption characteristics remotely without necessarily downloading all the data to the remote central monitoring station over the expensive wireless connection.

3. MineFleet Server:

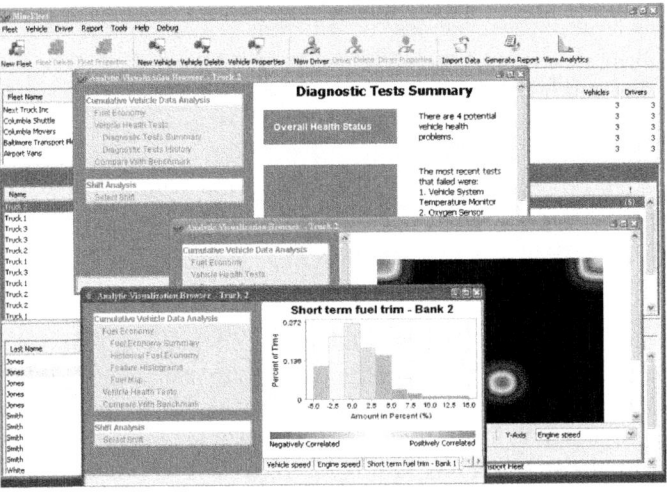

Fig. 3. User interface of the MineFleet Server. © Copyright, Agnik, LLC.

The MineFleet Server is in charge of receiving all the analytics from different vehicles, managing those analytics, and further processing them as appropriate. The MineFleet Server supports the following main operations: (i) interacting with the on-board module for remote management, monitoring, and mining of vehicle data streams and (ii) managing interaction with the MineFleet Web Services. It also offers a whole range of fleet-management related services that are not directly related to the main focus of this paper.

4. MineFleet Web Services: This module offers a web-browser-based interface for the MineFleet analytics. It also offers a rich class of API functions for accessing the MineFleet analytics which in turn can be integrated with third-party applications. Figure 4 shows one of the interfaces of the MineFleet Web Services.

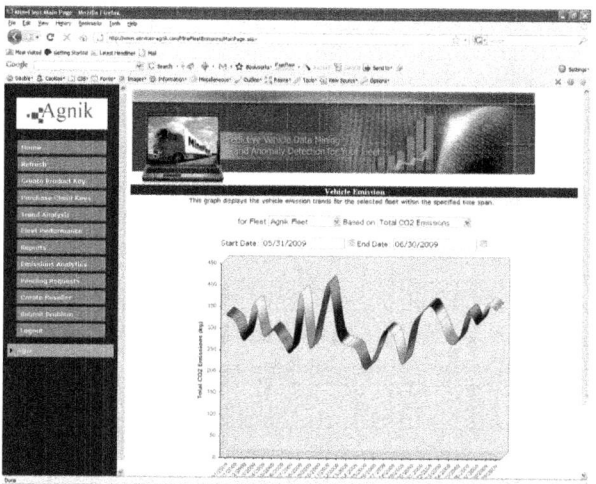

Fig. 4. User interface of the MineFleet Web Services. © Copyright, Agnik, LLC.

5. Privacy management module: This module plays an important role in the implementation of the privacy policies. This module manages the specific policies regarding what can be monitored and what cannot be.

In order to monitor the vehicle data streams using the on-board data management and mining module we need continuous computation of several statistics. For example, the MineFleet on-board system has a module that continuously monitors the spectral signature of the data which requires computation of covariance and correlation matrices on a regular basis. The on-board driving behavior characterization module requires frequent computation of similarity/distance matrices for data clustering and monitoring the operating regimes. Since the data are usually high dimensional, computation of the correlation matrices or distance (e.g. inner product, Euclidean) matrices is difficult to perform using their conventional algorithmic implementations.

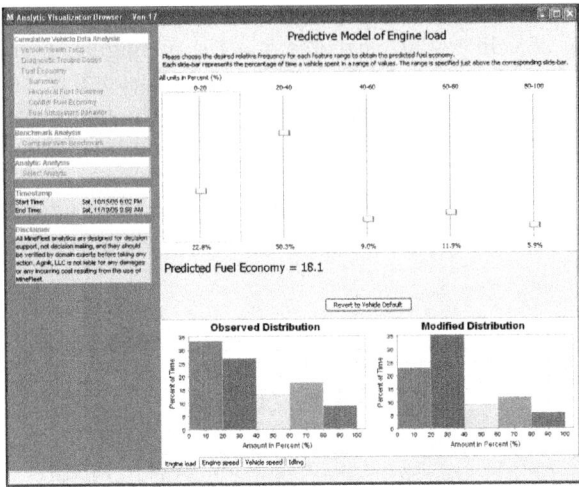

Fig. 5. Predictive fuel consumption analysis module of MineFleet. © Copyright, Agnik, LLC.

The incoming data sampling rate supported by the vehicle data bus limits the amount of time we get for processing the observed data. This usually means that we have only a few seconds to quickly analyze the data using the on-board hardware (e.g. the MF-DMP101 device). If our algorithms take more time than what we have in hand, we cannot catch up with the incoming data rate. In order to handle this situation, we need address the following issues:

1. We need fast "light-weight" techniques for computing and monitoring the correlation, covariance, inner product, and distance matrices that are frequently used in data stream mining applications.
2. We need algorithms that will do something useful when the running time is constrained. In other words, we allow the data mining algorithm to run for a fixed amount of time and expect it to return some meaningful information. For example, we give the correlation matrix computation algorithm certain number of CPU cycles for identifying the coefficients with magnitude greater than 0.7. If that time is not sufficient for computing all the correlation coefficients in the matrix then the algorithm should at least identify the portions of the matrix that may contain significant coefficients.

The rest of this paper deals with this problem of resource-constrained data mining specifically in the context of frequently computing sparse correlation matrices and monitoring changes in the correlation matrices computed from different windows of data streams. However, before discussing the developed algorithms, let us review the related material and discuss the problem of computing the correlation and distance matrices.

3   Frequent Computation of Correlation Matrices and Operating Regime Monitoring

Frequent computation of correlation, inner product, and distance matrices plays an important role in the performance of the MineFleet system. In this section, we offer a brief discussion on some of the MineFleet modules that require frequent computation of correlation matrices and monitoring changes in these matrices. First, let us consider the on-board predictive vehicle health monitoring module of MineFleet. This module is responsible for tracking the operating characteristics of the vehicle and detecting abnormal patterns from the vehicle health data.

Among other things, this module estimates the distribution of the data using different incremental parametric and non-parametric techniques. In this section, we discuss only one of them that makes use of an incremental operating regime identification technique. It identifies the safe operating regime of the vehicle in the low dimensional eigenspace of the covariance matrix by first clustering the data and then capturing the clusters using techniques from computational geometry. We assume that initially when the vehicle is certified to be in good health condition, we can observe its behavior, gradually generate the clusters, and then use the stable cluster descriptions to define the healthy operating regimes of the vehicle. Later, during the monitoring phase, the module notes whether or not the observed data point falls within the safe operating regime in the projected state space of the vehicle. If it does then everything may be okay; otherwise the module may raise a flag and reports unexpected behavior. The actual implementation of this module of MineFleet is more complex and it involves different statistical testing and confidence factor computation. However, we do not discuss those here since the focus of this paper is something else. Among other things, the performance-regime learning and monitoring module needs to monitor the changes in the correlation matrix and re-compute it, if necessary.

MineFleet also has a module for monitoring statistical dependency among different components of a vehicle system. Usually, the correlation matrix changes when something changes in the underlying functional relationship among the attributes. Such changes introduce non-stationary properties which are often an indicator of something unusual happening in the vehicle systems.

The following section discusses an algorithm to address this problem. Unlike the traditional correlation matrix computation approach, the algorithm presented in the following sections of this paper offer the following capabilities:

1. Quickly check whether or not the correlation matrix has changed using a probabilistic test.
2. Apply this test and a divide-and-conquer strategy to quickly identify the portions of the correlation matrix that contain the significantly changed coefficients.

In this paper, we primarily restrict ourselves to the change detection problem.

4   Correlation and Distance Matrices

The Pearson Product-Moment Correlation Coefficient or correlation coefficient for short is a measure of the degree of linear relationship between two random variables: a and b. The correlation between a and b is commonly defined as follows: $Corr(a, b) = \frac{Cov(a,b)}{\sigma_a \sigma_b}$, where $Cov(a, b)$ is the covariance between a and b; σ_a and σ_b are the standard deviations of a and b respectively. The correlation coefficient takes a value between -1 and +1. A correlation of +1 implies a perfect positive linear relationship between the variables. On the other hand, a correlation of -1 implies a perfect negative linear relationship between the variables. A zero correlation coefficient means the two variables vary independently. In this paper, we call a correlation coefficient significant if its magnitude is greater than or equal to a user given threshold.

In data mining applications we often estimate the correlation coefficient of a pair of features of a given data set. A data set comprised of features two X and Y, that has m observations, has m pairs (x_i, y_i) where x_i and y_i are the i-th observations of X and Y respectively. The following expression is commonly used for computing the correlation coefficient:

$$Corr(X,Y) = \frac{\sum x_i y_i - \frac{\sum x_i \sum y_i}{m}}{\sqrt{\left(\sum x_i^2 - \frac{(\sum x_i)^2}{m} \right) \left(\sum y_i^2 - \frac{(\sum y_i)^2}{m} \right)}}$$

If the data vectors have been normalized to have 0 mean and unit length (ℓ_2 norm), the resulting expression for correlation coefficient is a lot simpler.

$$Corr(X', Y') = \sum_{i=1}^{m} x_i y_i \tag{1}$$

In this paper, we assume that the data sets have been normalized first. Therefore, if U is the $m \times n$ data matrix with the m rows corresponding to different observations and the n columns corresponding to different attributes, the correlation matrix is a $n \times n$ matrix $U^T U$. This paper also uses the term correlation-difference matrix in the context of continuous data streams. If $Corr_t(X, Y)$ and $Corr_{t+1}(X, Y)$ are the correlation coefficients computed from the data blocks observed at time t and $t + 1$ respectively then the correlation-difference coefficient is defined as $|Corr_t(X, Y) - Corr_{t+1}(X, Y)|$. When there are more than two data columns corresponding to different attributes, we have a set of such coefficients that can be represented in the form of a matrix. This matrix will be called the correlation-difference matrix.

Also note that the problem of computing the Euclidean distance matrix is closely related to the correlation matrix and inner product computation problem. The Euclidean distance between the data vectors corresponding to X and Y, $\sum_{i=1}^{m}(x_i - y_i)^2 = \sum_{i=1}^{m}(x_i^2 + y_i^2 - 2x_i y_i) = \sum_{i=1}^{m} x_i^2 + \sum_{i=1}^{m} y_i^2 - 2\sum_{i=1}^{m} x_i y_i$.

The correlation coefficient computation is also very similar to the problem of computing the inner product [3] [9]. Therefore, in the rest of this paper we present the proposed algorithm only in the context of the correlation computation problem. An efficient technique for computing the correlation matrix is equally applicable to the inner product and Euclidean distance computation problem. These statistical computing primitives are directly useful for clustering, principal component analysis, and many other related statistical and data mining applications. Therefore, although the rest of the paper considers only the correlation matrix monitoring problem, the results have implications on solving many other related problems.

Efficient computation of the correlation matrix has been addressed in the literature. Zhu and Shasha exploited [10] an interpretation of the correlation coefficient as a measure of Euclidean distance between two data vectors in the Fourier representation. Computation of correlation coefficients in the Fourier domain is also frequently used in the signal processing literature. They developed the StatStream system which has been applied to compute correlation matrices from continuous data streams. Their results show scalable improved performance compared to the naive way to compute the correlation coefficients. This technique is designed for desktop applications and the overhead makes it unsuitable for the resource-constrained environments where monitoring of the correlation matrix is the primary objective. Alqallaf et al. [2] considered the problem of robust estimation of the covariance matrix for data mining applications. These techniques are designed for desktop applications and the computational overhead is just not appropriate for the embedded environment considered here in this paper. The following section briefly discusses an efficient algorithm to detect changes in the correlation matrix.

5 Divide and Conquer Approach towards Correlation Matrix Monitoring

Given an $m \times n$ data matrix U with m observations and n features, the correlation matrix is computed by $U^T U$ assuming that the columns of U are normalized to have zero mean and unit variance. We are particularly interested in sparse correlation matrices because of the monitoring application we have in mind. However, such correlation matrices are also widely prevalant since in most real-life high dimensional applications features are not highly correlated with every other feature. Instead only a small group of features are usually highly correlated with each other. This results in a sparse correlation matrix. In most stream applications, including the MineFleet, the difference in the consecutive correlation matrices generated from two subsequent sets of observations is usually small, thereby making the difference matrix a very sparse one.

A straight forward approach to compute the correlation matrix using matrix multiplication takes $O(mn^2)$ multiplications. The objective of this section is to present an efficient technique for computing sparse correlation matrices. If the matrix has $O(c)$ number of significant coefficients then the algorithm runs in $O(c \log n)$ time. In order to achieve this, we first demonstrate how we can estimate the sum of squared values of the elements in the correlation matrix that are above the diagonal. We define this sum as $C = \sum_{1 \leq j_1 < j_2 \leq n} Corr^2(j_1, j_2)$, where $Corr(j_1, j_2) = \sum_{i=1}^{m} u_{i,j_1} u_{i,j_2}$ represents the correlation coefficient between j_1–th and j_2–th columns of the data matrix U. We can estimate C using an approach that is similar in spirit to that of [1]. First consider the following lemma.

Lemma 1 ([6]). *Consider an $m \times n$ data matrix U and a n-dimensional random vector $\sigma = [\sigma_1, \sigma_2, \cdots, \sigma_n]^T$, where each $\sigma_j \in \{-1, 1\}$ is independently and identically distributed. Let v_i be a random projection of the i-th row of the data matrix U using this random vector σ, i.e.*

$$v_i = \sum_{j=1}^{n} u_{i,j} \sigma_j$$

Define a random variable $Z = (\sum_{i=1}^{m} v_i^2 - n)/2$ and $X = Z^2$. Then,

$$E[X] = \sum_{1 \leq j_1 < j_2 \leq n} Corr^2(j_i, j_2) = C \tag{2}$$

and

$$Var[X] \leq 2C^2.$$

where $E[X]$ and $Var[X]$ represent the expectation and the variance of the random variable X, respectively.

Lemma 1 provides results that lay the foundation of a probabilistic technique to estimate $\sum_{1 \leq j_1 < j_2 \leq n} Corr^2(j_i, j_2)$ which can be used for testing the existence of a significant correlation coefficient.

Consider a set of $s = s_1 s_2$ random σ vectors that are from i.i.d. or 8–wise independent. Corresponding to each of the σ vectors, we maintain random variables $X_{i,j}$ using the data matrix. Let $Y_1, Y_2, \cdots, Y_{s_2}$ be such that Y_i is the mean of $X_{i,j} : 1 \leq j \leq s_1$. Let Y be the median of the Y_i's. We output the median Y as the estimate for C.

Let $s_1 = 16/\lambda^2$ and $s_2 = 2\lg(1/\epsilon)$. By Chebyshev's inequality, for each fixed p, $1 \leq p \leq s_2$,

$$Prob[|Y_p - C| > \lambda C] \leq \frac{Var[Y_p]}{\lambda^2 C^2} \leq \frac{2C^2}{s_1 \lambda^2 C^2} = \frac{1}{8}.$$

By considering the median of s_2 of such random variables, we have by the standard Chernoff bounds that the probability that the estimate deviates from C by more than λC is at most ϵ. The approach can be further simplified by considering a sample of k columns of the data matrix for computing each of the $X_{i,j}$'s to estimate C.

Next, we describe how this approach can be used to identify the significant coefficients of the correlation matrix. Given a set $L = \{j_1, j_2, \cdots, j_k\}$ of k–columns from the data matrix U, let U_L be the data matrix with only data column vectors from the set L, i.e. $U_L = [u_{j_1}, u_{j_2}, \cdots u_{j_k}]$, in order to detect if any of these columns are strongly correlated we first estimate C for U_L using the above approach. Let C_L be the true value of C over this pruned dataset and Y_L be the estimated value. If any of the correlation coefficients has a magnitude greater than θ then the true value of C_L must be a value greater than of equal to θ^2. We use this test to determine whether or not there are any significant correlations among the data columns in U_L. If the estimated value Y_L is less than θ^2, we declare that the columns $L = \{j_1, j_2, \cdots, j_k\}$ are not significantly correlated.

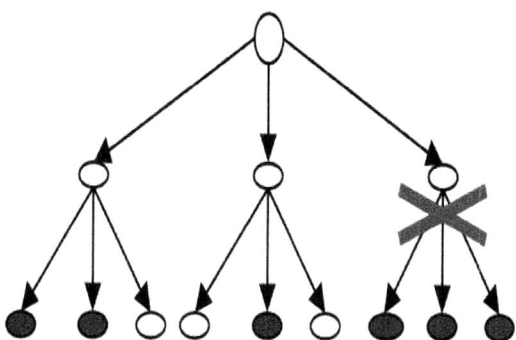

Fig. 6. A graphical representation of the divide-and-conquer-strategy-based algorithm for computing the significant correlation coefficients

This way, the above technique can be used to design a tree-based divide and conquer strategy that first checks the possible existence of any significant correlation coefficient among a set of data columns before actually checking out every pair-wise coefficient. If the test turns out to be negative then we discard the corresponding correlation coefficients for further consideration. Figure 6 shows the intuitive idea behind this Fast Monitoring of Correlation matrix (FMC) algorithm. The algorithm performs a tree-search in the space of all correlation coefficients. Every leaf-node of this tree is associated with a unique coefficient; every internal node a is associated with the set of all coefficients corresponding to the leaf-nodes in the subtree rooted at node a. The algorithm tests to see if the estimated $C_a \geq \theta^2$ at every node starting from the root of the tree. If the test determines that the subtree is not expected to contain any significant coefficient then the corresponding sub-tree is discarded. Search proceeds in that sub-tree otherwise.

At a node with L features the test-of-significance involves computation of v_i's for $i = 1, \cdots, m$. It can be shown that the time to perform a test at a node with $|L|$ attributes is the time taken for $O(s_1|L|m)$ additions, $O(s_1 m)$ multiplications and $O(s_2 \log s_2)$ comparisions. Note that this is much faster than the naive approach that requires $O(m|L|^2)$ additions and multiplications.

For a correlation matrix with $O(c)$ number of significant coefficients, the cost of the FMC algorithm in terms of multiplications is $O(c \log n)O(s_1 m) = O(s_1 cm \log n)$. However, the number of additions carried out at a node depends on the number features at that node. In fact this is $O(s_1 m|L|)$ additions at a node with $|L|$ features. Therefore nodes closer to the root have more number of features and hence costs more number of additions. However nodes with large number of features are few in number, thus balancing out the number of additions across nodes along a path to the significant coefficient. More precisely, the number of additions that the FMC algorithm performs is $O(cs_1 nm)$.

The FMC algorithm has a space requirement of $O(c \log n)$ to store the significance-test results at each of the nodes where the tests are performed. In addition we should also consider the storage requirement for storing the seeds and generating the random σ's for each of the $s_1 s_2$ Z's at any node.

Note that the standard matrix-multiplication-based technique for computing the correlation matrix requires $O(n^2)$ space and $O(mn^2)$ additions and $O(mn^2)$ multiplications.

The following section extends the FMC algorithm to the stream monitoring scenario which is the main focus of the current application.

6   Detecting Changes in the Correlation Matrices

This section extends the sparse-correlation matrix computation technique to a stream data environment for monitoring the correlation difference matrices, the targetted application for the research reported here.

Consider a multi-attribute data stream scenario where each time stamp is associated with a window of observations from the stream data. More specifically, let $U^{(t)}$ and $U^{(t+1)}$ be the consecutive data blocks at time t and $t+1$ respectively. Let $Corr(j_1^{(t)}, j_2^{(t)})$ be the correlation coefficients between j_1–th column and j_2–th column of $U^{(t)}$ and similarly let $Corr(j_1^{(t+1)}, j_2^{(t+1)})$ be the correlation coefficients between j_1–th column and j_2–th column of $U^{(t+1)}$. Along the same lines, let $Z^{(t)}$ and $Z^{(t+1)}$ be the estimated values of Z for the two data blocks at time t and $t+1$ respectively. Let $X^{(t)}$ and $X^{(t+1)}$ be the corresponding estimated values of X. Note that we use the same σ's for computing $X^{(t)}$ as well as $X^{(t+1)}$. Let us define,

$$\Delta^{(t+1)} = Z^{(t+1)} - Z^{(t)}$$

Now we can write,

$$\Delta^{(t+1)} = \sum_{1 \le j_1 < j_2 \le n} \sigma_{j_1} \sigma_{j_2} Corr(j_i^{(t+1)}, j_2^{(t+1)})$$

$$- \sum_{1 \le j_1 < j_2 \le n} \sigma_{j_1} \sigma_{j_2} Corr(j_i^{(t)}, j_2^{(t)})$$

$$= \sum_{1 \le j_1 < j_2 \le n} \sigma_{j_1} \sigma_{j_2} \left(Corr(j_i^{(t+1)}, j_2^{(t+1)}) - Corr(j_i^{(t)}, j_2^{(t)}) \right)$$

Then we can find the expected value of $\left(\Delta^{(t+1)}\right)^2$ in a manner similar to finding $E[X]$ described earlier.

$$E[\left(\Delta^{(t+1)}\right)^2] =$$

$$\sum_{1 \leq j_1 < j_2 \leq n} \left(Corr(j_i^{(t+1)}, j_2^{(t+1)}) - Corr(j_i^{(t)}, j_2^{(t)})\right)^2$$

This can be used to directly look for significant changes in the correlation matrix. We should note that the difference correlation matrix (i.e. the changes in the matrix) is usually very sparse since most of the time vehicle systems do not perform unusually; rather they work following well understood principles of mechanical and electrical systems. The following section presents experimental results documenting the performance of the proposed algorithm in the context of the MineFleet application.

7   Experimental Results

This section presents the experimental results documenting the performance of the FMC algorithm using vehicle data streams. We consider the problem of monitoring the correlation matrices computed from different data windows sampled from the data streams. Our objective is to compare the performance of the naive enumerative algorithm and the FMC in correctly detecting the following scenarios:

1. No changes in the correlation matrix over two consecutive data windows sampled from the streams.
2. No significant changes in the correlation matrix. Note that this is different from the previous scenario since in this case the correlation matrices do not stay invariant although the changes are insignificant with respect to a given threshold.
3. Detecting significant changes in the correlation matrix.

The following sections consider each of these scenarios.

7.1   Detecting No Changes

In this section our objective is to study the performance of the FMC algorithm when the correlation matrix stays invariant. We performed several experiments using data stream windows that produce the same correlation matrices. FMC works very well, as suggested by the analytical results. It always identifies no change by performing the very first test at the root node of the tree. As a result the running time is constant compared to the quadratic order running time for the naive enumerative approach. As figure 7 shows, the performance of FMC is significantly better than that of the naive approach.

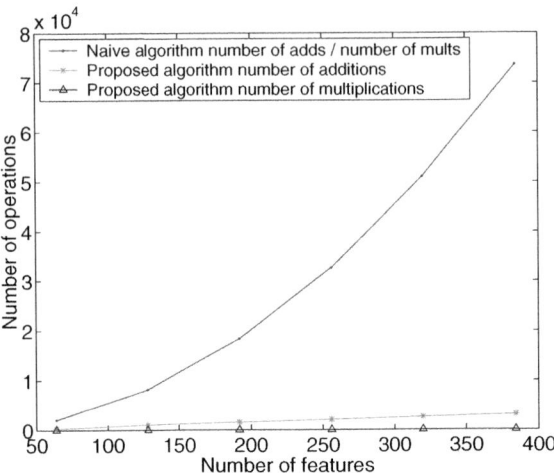

Fig. 7. Number of multiplications and additions performed by the enumerative algorithm and FMC for correctly detecting no changes in the correlation matrix

7.2   Detecting No Significant Changes

This section considers the scenario where the correlation matrices are slightly different, resulting in a correlation-difference matrix that is not a null matrix but it does not contain any significantly changed (with respect to the given threshold) coefficients either. Table 1 shows the performance of the proposed algorithm for overlapping windows from the stream with insignificant but non-zero changes. The naive enumerative algorithm requires (2016 × number-of-data-rows) multiplications and additions. FMC detects no significant changes with approximately half the number of additions and a very small fraction of multiplications. The FMC algorithm clearly outperforms the enumerative algorithm on this ground.

7.3   Detecting Significant Changes

This section considers the problem of detecting significant changes when some of the coefficients in the correlation matrix have changed beyond the given threshold. In this situation, the algorithm has the two following goals:

1. Detect that something has indeed changed in the correlation matrix
2. Identify the portions of the matrix that are likely to contain the significantly changed coefficients.

This section first reports the results of experiments with overlapping windows of data where the difference-correlation matrix contains exactly 6 significant entries and the magnitude of the difference is greater than 0.6.

Table 1. Number of multiplications and additions (with a scaling factor of number-of-data-rows) performed by FMC for correctly detecting no significant changes in the correlation matrix. Threshold value is 0.6.

	Number of Additions		Number of Multiplications		Number of Nodes	
r	μ	σ	μ	σ	μ	σ
2	875.6	564.7	10.4	7.7	5.2	3.8
4	722.4	468.2	3.4	2.6	1.8	1.3
6	1692.6	1376.9	6	6.2	3	3.1
8	1025	0	2	0	1	0
10	1281	0	2	0	1	0

Table 2. Number of multiplications and additions (with a scaling factor of number-of-data-rows) performed by the naive algorithm and FMC for correctly detecting significant changes in the correlation matrix and identifying the portions of the matrix with the significantly changed coefficients

	Number of Significant Coefficients Detected		Number of Multiplications		Number of Additions	
r	μ	σ	μ	σ	μ	σ
2	4.8	1.09	12.8	4.3	1081.6	282.6
4	5.2	1.09	16	0	2456	250.4
8	4.8	1.09	14.4	3.5	4717.6	917.6
12	4.8	0.89	14.4	3.5	7072.8	1833.5
16	4.4	0.89	14.4	3.5	9428	1833.5
20	4.4	0.89	16	0	128008	0

In all the experiments reported here, FMC returns the correct answer for the first problem (1) listed above. Our experiments were carried out with finite resource constraints. Table 2 shows the number of multiplications and additions when the algorithm is allowed to explore only 8 nodes in the tree. Even with this restriction on computation, the algorithm could detect the regions of the correlation matrix with most of the significantly changed coefficients. Figure 8 shows the 64×64 dimensional correlation-difference matrix. Since the matrix is symmetric, the matrix is divided into two different regions with different color shadings. The right-upper triangle shows four different regions (with different color shades) corresponding to the regimes defined by the nodes of the tree constructed by the algorithm. Six bright dots in the right-upper triangle correspond to the significant entries. The darker upper-rightmost region is correctly

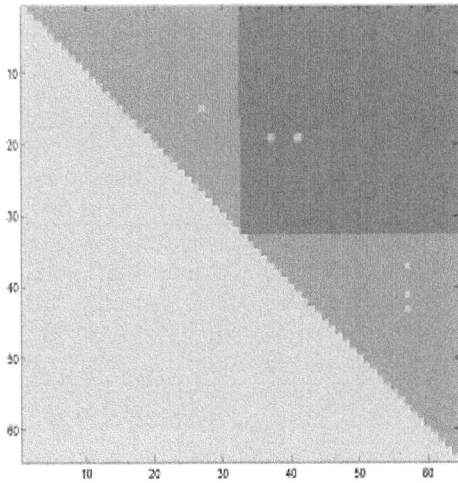

Fig. 8. This figure shows the 64 × 64 dimensional correlation-difference matrix. Six bright dots represent the significant entries.

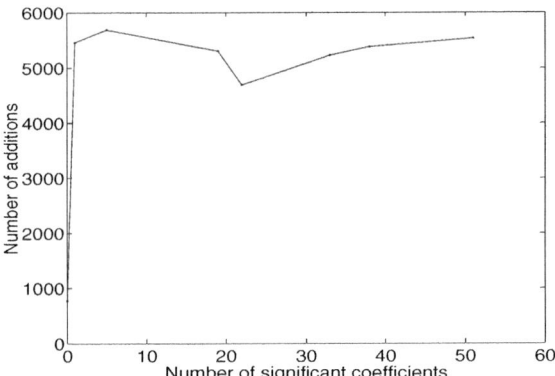

Fig. 9. Average number of additions performed by the proposed algorithm (scaled by the number of rows) vs. the number of significant changes in the correlation matrix

discarded by the algorithm since it does not contain any significant entry. Note that all the six significant coefficients are covered by the nodes selected by the FMC algorithm.

Next we examine the performance of the algorithm as the number of significant changes in the correlation matrix varies. We consider vehicle data collected from a 2003 Ford Taurus. This dataset contains 831 rows and we will use 64 of its

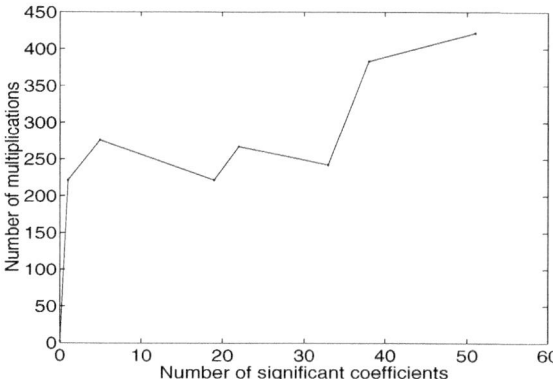

Fig. 10. Average number of multiplications performed by the proposed algorithm (scaled by the number of rows) vs. the number of significant changes in the correlation matrix

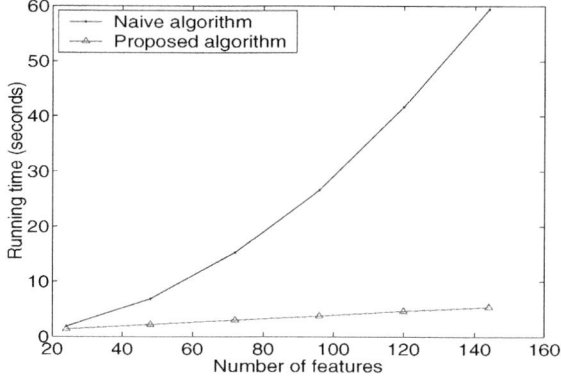

Fig. 11. Comparison of the running time of the proposed and the naive algorithm for detecting no changes in the correlation matrix

columns. We create two blocks of data, each containing the 831 rows of the Ford Taurus data. We make the correlation matrices between the two blocks different by altering some of the columns in the first block. This is performed for introducing a controlled amount of significant changes in the correlation matrices. After each alteration we run our algorithm and record the average number of additions and multiplications required as it searches for differences between the two correlation matrices corresponding to the two blocks of data. In this experiment we fix the number of random vectors to $r = 8$ and consider a coefficient significant if its magnitude is greater than 0.6. We perform these experiments using finite resource constraints in order to quickly identify regions (subtrees of the overall search tree used by FMC) of the correlation difference matrix that contain significantly changed coefficients without performing an exhaustive search of these regions for exactly identifying the changed coefficients.

Figure 9 and Figure 10 present the results of the experiment described above. The traditional correlation matrix computation requires (2016×number of rows) additions and multiplications. The FMC algorithm requires a fairly large number of additions, but it requires considerably fewer multiplications than the naive approach. Due to the finite resource constraints, the number of additions does not increase drastically as the number of significant coefficients increases. The bounds on the numbers of additions and multiplications is fixed a priori. The goal is to identify the regimes (i.e. the sub-trees) of the significantly changed coefficients as precisely as possible. Results clearly show that the algorithm is able to locate these regions using fewer multiplications then required by the naive enumerative approach and a modest, relatively fixed number of additions.

8   Conclusions

This paper presented a brief overview of MineFleet, a real-time vehicle data stream mining and monitoring system. MineFleet is one among the first distributed data mining systems that is designed for mobile applications. We believe that this technology will find many other applications in different domains where monitoring time-critical data streams is important and central collection of data is an expensive proposition.

The paper mainly focused on a particular aspect of the MineFleet system—monitoring correlation and distance matrices. It offered a probabilisitc technique for efficiently detecting changes in the correlation matrix and identifying portions of the matrix that are likely to contain the significantly changed coefficients. MineFleet contains a proprietary version of this algorithm and it plays a critical role in the real-time performance of the vehicle on-board module for data analysis.

The proposed technique adopts a divide-and-conquer strategy that makes use of a test to check whether or not a subset of correlation coefficients contains any significant coefficient. The test allows us to prune out those subsets of coefficients that do not appear to contain any significant one. The technique is particularly suitable for efficiently monitoring changes in coefficient matrices and computing large sparse correlation matrices. The proposed algorithm really made a difference in the performance of the MineFleet system. For example Figure 11 shows the comparative running time of the proposed and the naive algorithm on a Dell Axim PDA. The experiments are performed for increasing number of features. The proposed algorithm detects no changes in the correlation matrix at a very minimal cost saving several seconds of clock time. A few seconds of saving in running time is a major achievement in a resource constrained environment like what the on-board module of MineFleet uses.

Our preliminary experimental results with controlled and real-life vehicle stream data appear promising. However, there are several issues that need to be addressed in order to make the performance of the proposed algorithm more attractive. As we noted earlier, the accuracy of the algorithm depends on the value of r, i.e. number of different randomized trials. For relatively larger values of r, the accuracy is usually excellent; however, the running time goes up accordingly. Therefore, it will be nice if we can construct a deterministic version of the test that does not require multiple trials at every node of the tree. The performance of the algorithm also depends on the overhead necessary for maintaining and manipulating the tree structure. Therefore, paying attention to the systems issues is important, particularly for the run-time performance on-board an embedded device. The current implementation of the technique is primarily designed for monitoring changes in the correlation matrix. Once we identify that the matrix has significantly changed it is usually better to use the naive correlation coefficient computation technique to generate the new matrix exactly. The algorithm also seems to work well when the matrix is not significantly changing frequently. In general, if the application requires continuous low-overhead monitoring of occasionally changing correlation or distance matrices then FMC appears to be a quite appropriate choice.

References

1. Alon, N., Matias, Y., Szegedy, M.: The space complexity of approximating the frequency moments. In: Proceedings of the ACM Symposium on Theory of Computing, pp. 20–29 (1996)
2. Alqallaf, F., Konis, K., Martin, R., Zamar, R.: Scalable robust covariance and correlation estimates for data mining. In: ACM Press (ed.) Proceedings of the Eighth ACM SIGKDD International Conference on Knowledge Discovery and Data Mining, pp. 14–23 (2002)
3. Falk, R., Well, A.: Many faces of the correlation coefficient. Journal of Statistics Education 5(3) (1997)
4. Kargupta, H., Bhargava, R., Liu, K., Powers, M., Blair, P., Bushra, S., Dull, J., Sarkar, K., Klein, M., Vasa, M., Handy, D.: Vedas: A mobile and distributed data stream mining system for real-time vehicle monitoring. In: Proceedings of the SIAM International Data Mining Conference, Orlando (2004)
5. Kargupta, H., Chan, P.: Advances in Distributed and Parallel Knowledge Discovery. AAAI/MIT Press (2000)
6. Kargupta, H., Puttagunta, V., Klein, M.: On-board vehicle data stream monitoring using minefleet and fast resource constrained monitoring of correlation matrices. Special issue of New Generation Computing Journal on Learning from Data Streams 25(1), 5–32 (2007)
7. Kargupta, H., Sivakumar, K.: Existential pleasures of distributed data mining. In: Next Generation Data Mining: Future Directions and Challenges. MIT/AAAI Press (2004)

8. Srivastava, A.N., Stroeve, J.: Onboard detection of snow, ice, clouds and other geophysical processes using kernel methods. In: Proceedings of the ICML 2003 Workshop on Machine Learning Technologies for Autonomous Space Sciences (2003)
9. Weldon, K.L.: A simplified introduction to correlation and regression. Journal of Statistics Education 8(3) (2000)
10. Zue, Y., Shasha, D.: Statistical monitoring of thousands of data streams in real time. In: Proceedings of the 28th VLDB Conference, Hong Kong, China (2002)

Author Index